W9-CCU-637

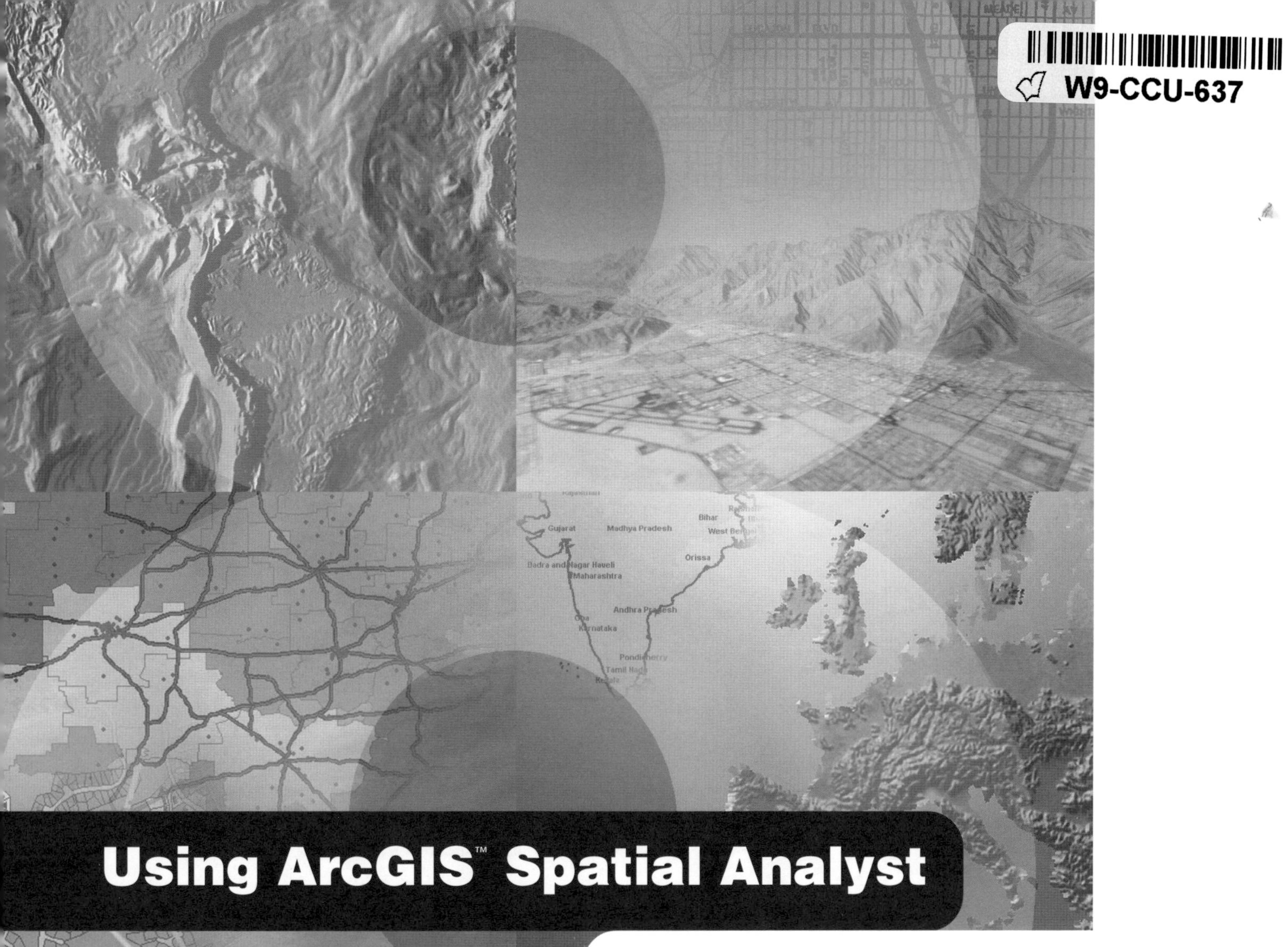

Using ArcGIS™ Spatial Analyst

GIS by ESRI™

Jill McCoy and Kevin Johnston

Copyright © 2001–2002 ESRI
All rights reserved.
Printed in the United States of America.

The information contained in this document is the exclusive property of ESRI. This work is protected under United States copyright law and other international copyright treaties and conventions. No part of this work may be reproduced or transmitted in any form or by any means, electronic or mechanical, including photocopying and recording, or by any information storage or retrieval system, except as expressly permitted in writing by ESRI. All requests should be sent to Attention: Contracts Manager, ESRI, 380 New York Street, Redlands, CA 92373-8100, USA.

The information contained in this document is subject to change without notice.

DATA CREDITS

Yellowstone National Park data: National Park Service, Yellowstone National Park, Wyoming

Joshua Tree National Park data: National Park Service, Department of the Interior, U.S. Government

Haul Cost Analysis map: Boise Cascade Corporation, Boise, Idaho

Quick-start tutorial data: courtesy of the State of Vermont

CONTRIBUTING WRITERS

Steve Kopp, Brett Borup, Jason Willison, Bruce Payne

U.S. GOVERNMENT RESTRICTED/LIMITED RIGHTS

Any software, documentation, and/or data delivered hereunder is subject to the terms of the License Agreement. In no event shall the U.S. Government acquire greater than RESTRICTED/LIMITED RIGHTS. At a minimum, use, duplication, or disclosure by the U.S. Government is subject to restrictions as set forth in FAR §52.227-14 Alternates I, II, and III (JUN 1987); FAR §52.227-19 (JUN 1987) and/or FAR §12.211/12.212 (Commercial Technical Data/Computer Software); and DFARS §252.227-7015 (NOV 1995) (Technical Data) and/or DFARS §227.7202 (Computer Software), as applicable. Contractor/Manufacturer is ESRI, 380 New York Street, Redlands, CA 92373-8100, USA.

ESRI and the ESRI globe logo are trademarks of ESRI, registered in the United States and certain other countries; registration is pending in the European Community. ArcMap, ArcCatalog, ArcGIS, and GIS by ESRI are trademarks and www.esri.com and arconline.esri.com are service marks of ESRI. Microsoft is a registered trademark and the Microsoft Internet Explorer logo is a trademark of Microsoft Corporation. HP and LaserJet are registered trademarks of Hewlett–Packard.

Other companies and products mentioned herein are trademarks or registered trademarks of their respective trademark owners.

Contents

Getting started

Understanding rasters and analysis

Performing analysis

Getting started

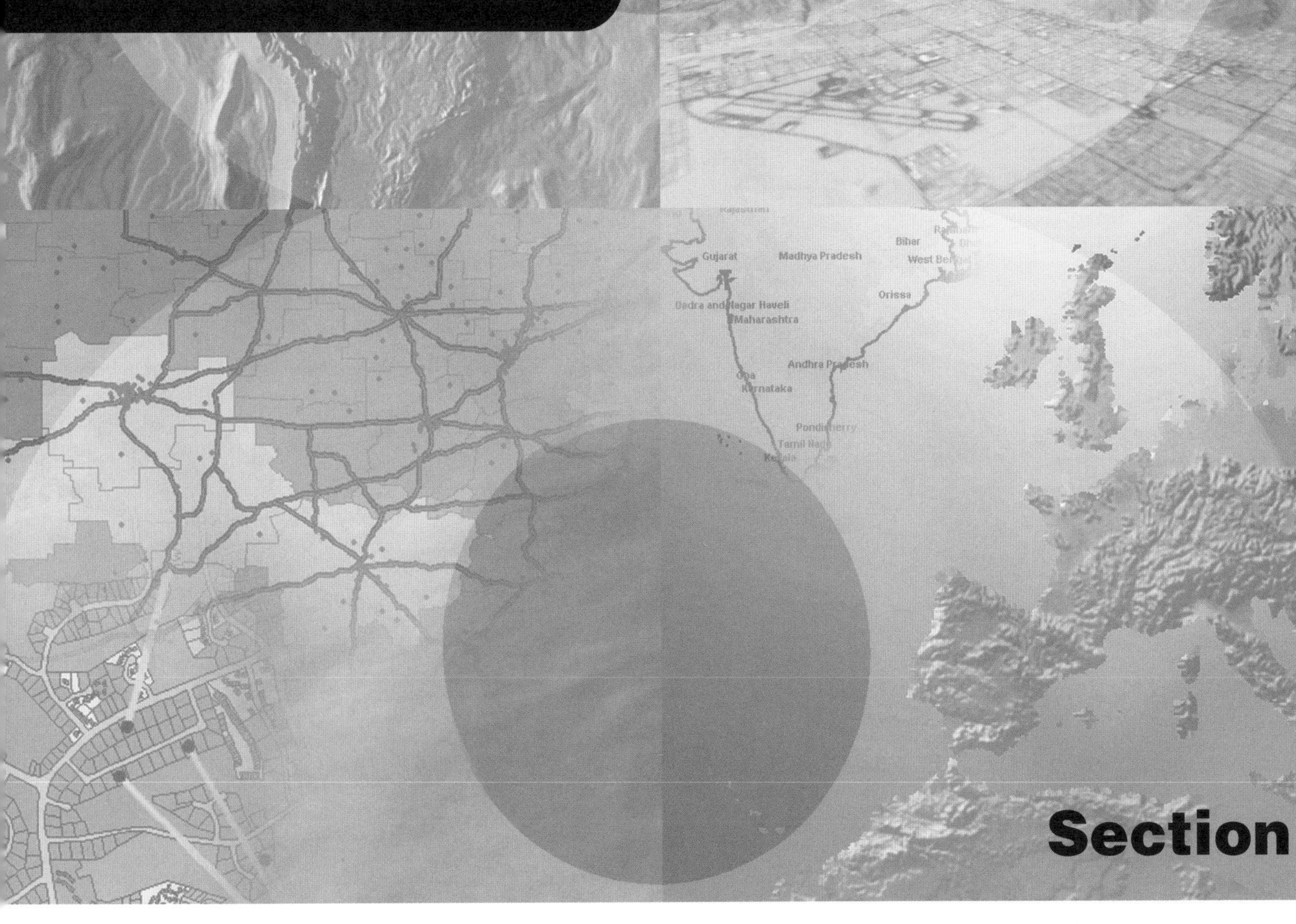

Section 1

Introducing ArcGIS Spatial Analyst

1

IN THIS CHAPTER

- **Deriving information from data**
- **Identifying spatial relationships**
- **Finding suitable locations**
- **Calculating cost of travel**
- **Tips on learning ArcGIS Spatial Analyst**

A key benefit of geographic information systems (*GIS*) is the ability to apply spatial operators to GIS data to derive new information. These tools form the foundation for all spatial modeling and geoprocessing. Of the three main types of GIS data—raster, vector, and tin—the raster data structure provides the richest modeling environment and operators for spatial analysis. ESRI® ArcGIS™ Spatial Analyst extension adds a comprehensive, wide range of cell-based GIS operators to ArcGIS.

- **Derive new information.** Apply Spatial Analyst tools to create useful information—watershed delineation, surface estimation, and classification—for example, derive distance from roads or calculate population density.
- **Identify spatial relationships.** Explore relationships between layers through weighted overlay and combinations. Spatial Analyst contains a rich set of Map Algebra tools for cell-based modeling.
- **Find suitable locations.** By combining layers, find areas that are the most suitable for particular objectives (e.g., siting a new building, or analyzing high risk areas for flooding or landslides).
- **Calculate travel cost.** Create travel cost surfaces to identify optimum corridors. Factor in economic, environmental, and other objectives.
- **Work with all cell-based GIS data.** Regardless of the raster format, Spatial Analyst allows you to combine them in your analysis.

These operations and much more are possible. As a GIS modeler, this is the central toolset you'll use for analysis and modeling. The next few pages will introduce you to what is possible with ArcGIS Spatial Analyst.

Deriving information from data

Using Spatial Analyst functions you can create a rich set of informative maps from your data. Create a hillshade to use as a backdrop of the terrain to support other data layers. Calculate slope, aspect, and contours, or create a map displaying visibility. Use derived data together to help solve spatial problems.

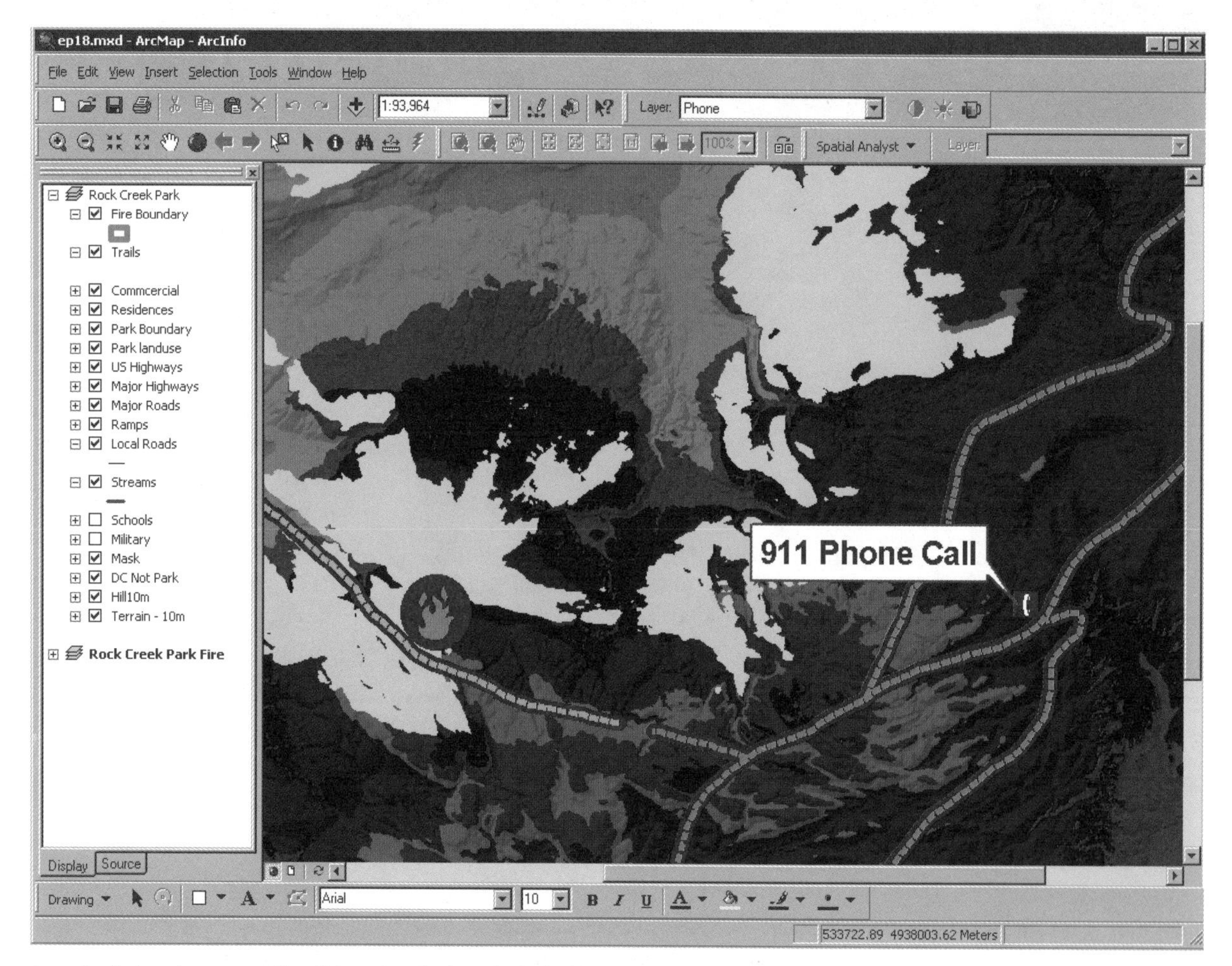

In order to break a suspect's alibi, a viewshed analysis finds out if he actually would have been able to see the location of the fire from where he called it in, claiming he saw flames. Areas drawn in yellow identify the locations from which the fire would have been seen. This visibility analysis demonstrates that he could not have seen the flames from the phone booth.

Identifying spatial relationships

Spatial Analyst provides tools to model spatial relationships.

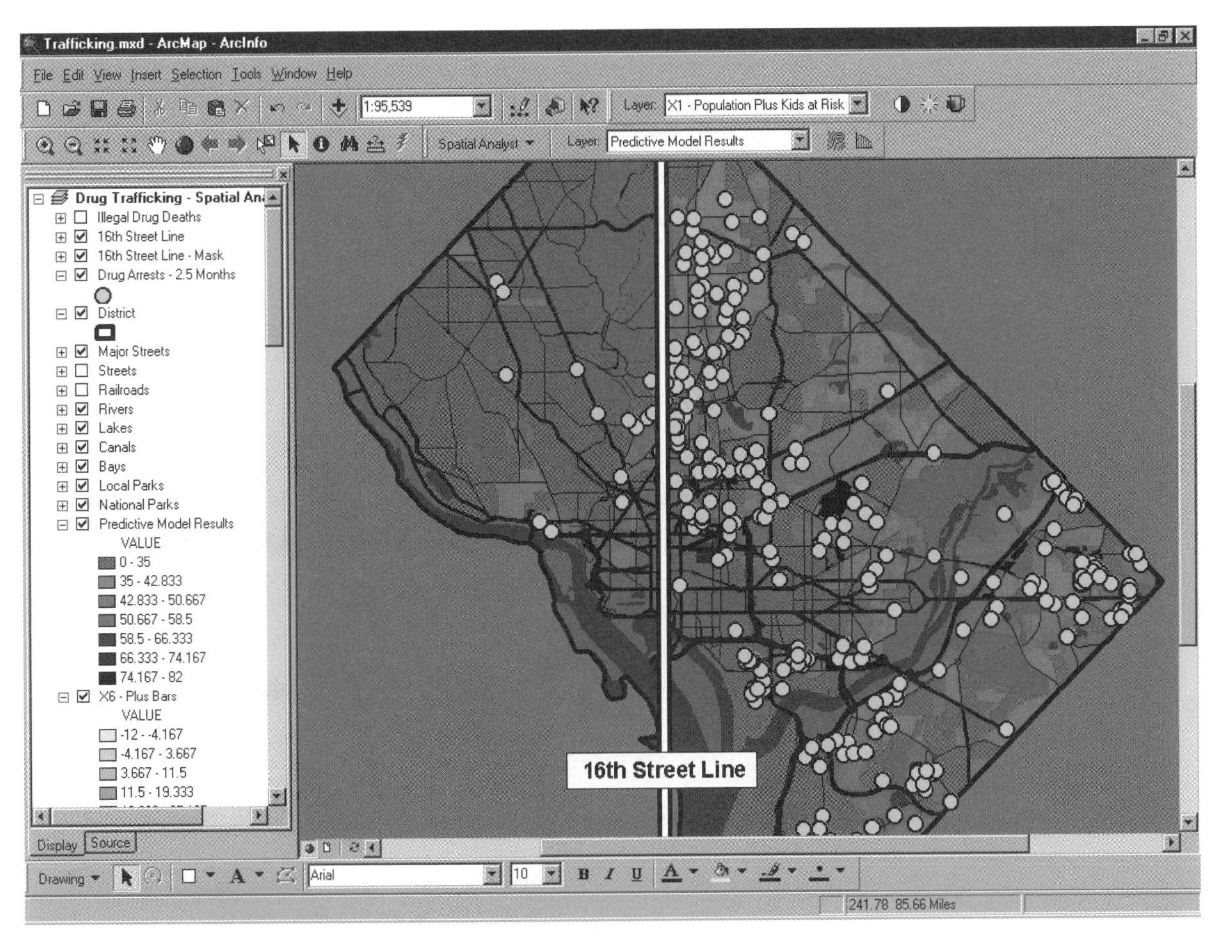

Model results aid in visual analysis. The darker red areas show locations predicting the highest level of drug traffic while the yellow dots represent the drug arrest locations for a three-month period. There is a high correlation between the two. There is also a marked difference in the number of arrests when you go west of 16th Street.

Finding suitable locations

Use the Spatial Analyst to query your data to identify locations that meet your set of objectives or produce a suitability map, combining datasets to analyze suitability.

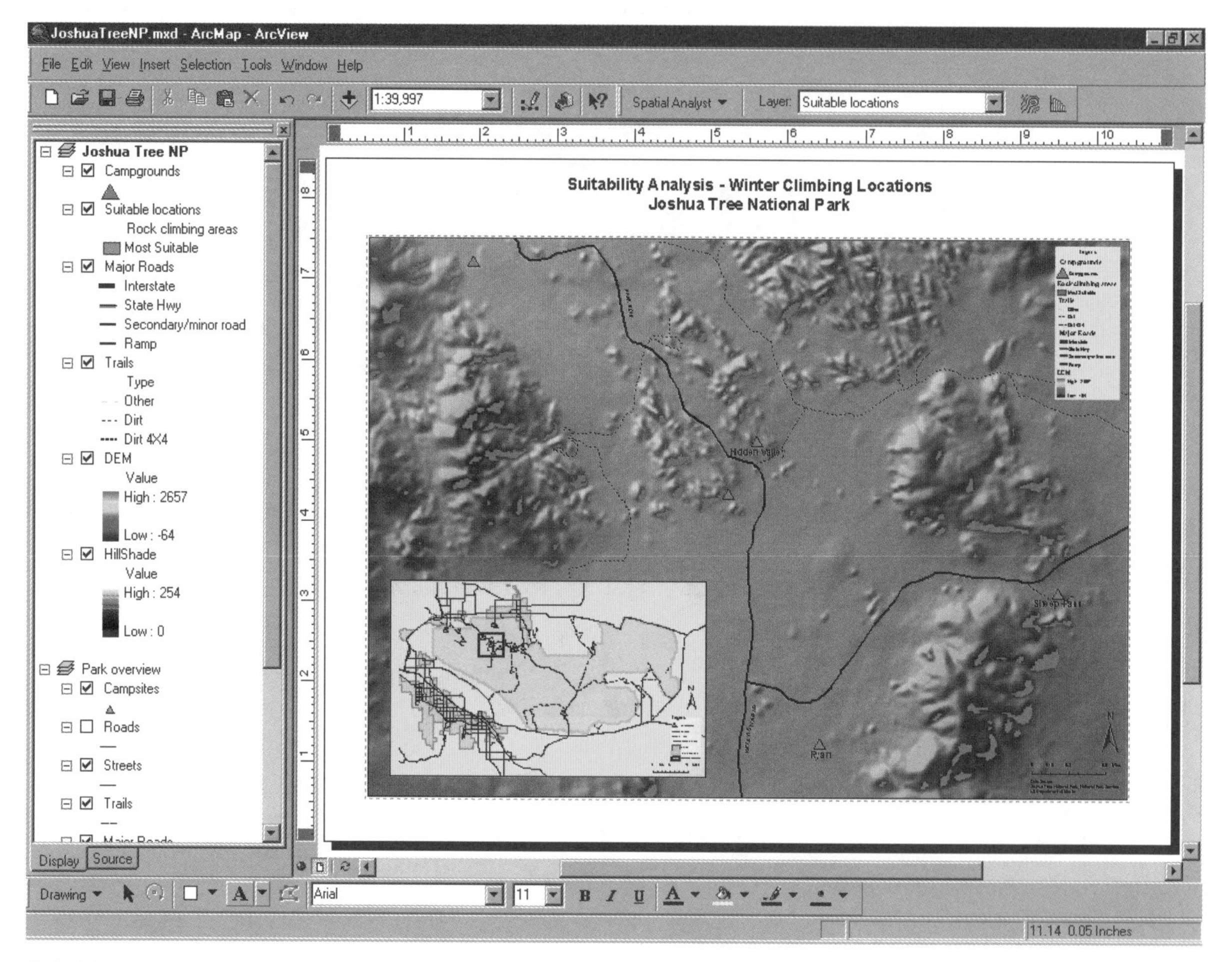

Suitable locations for winter rock climbing, based on distance from a campsite and steep, south-facing slopes

Calculating cost of travel

Travel cost analysis involves modeling to generate the cost surface and then calculating optimum corridors across the surface. Calculating travel cost can provide a rich set of information for decision making.

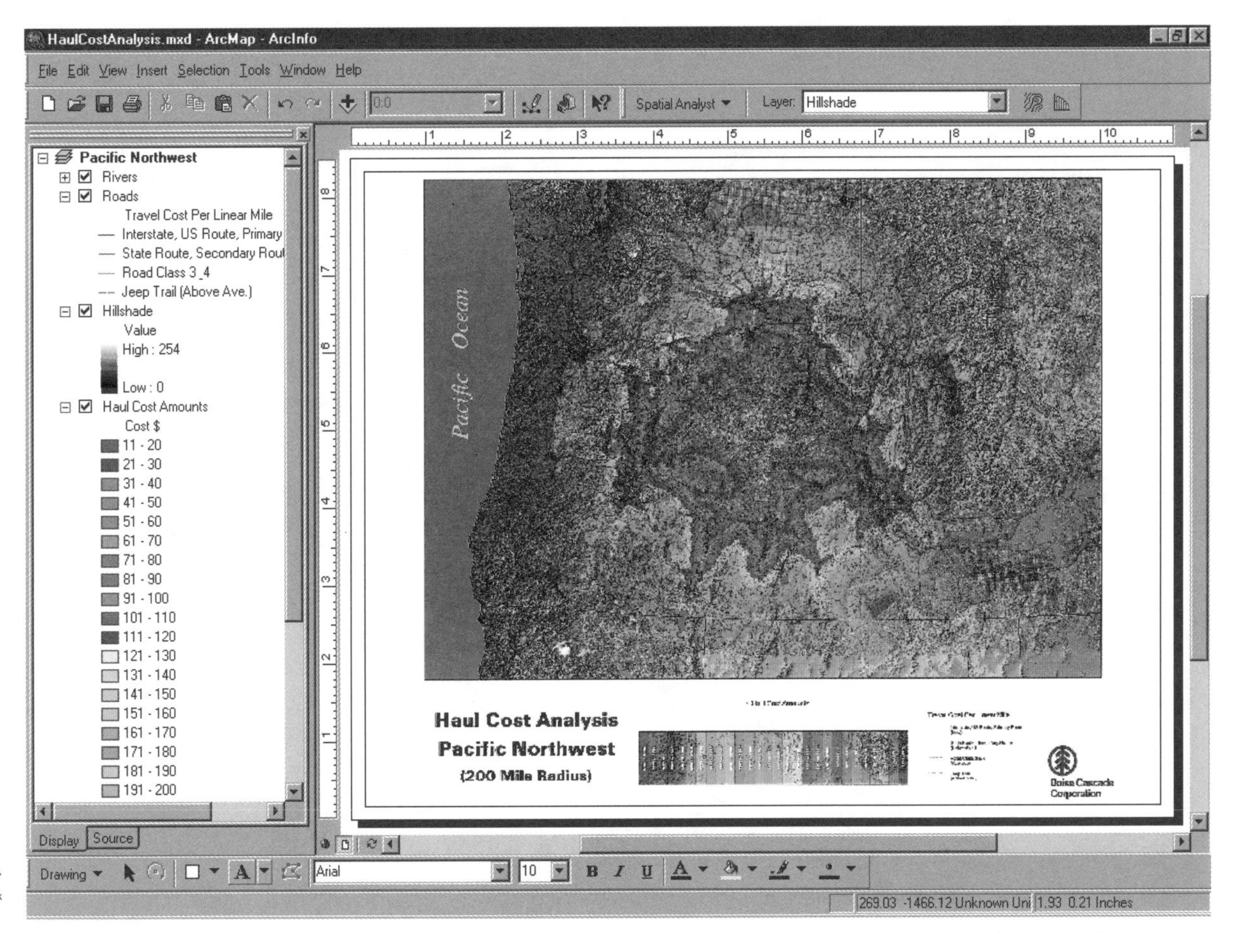

Haul Cost Analysis
Boise Cascade Corporation,
Boise, Idaho
Brian Liberty, Nick Blacklock
Copyright @ 1997

This map displays the least-cost travel for timber transport within a 200-mile radius of each mill. It considers obstacles to travel and estimates the cost in dollars to transport wood from each location to the nearest mill.

Tips on learning Spatial Analyst

If you're new to the concept of geographic information systems (GIS), remember that you don't have to know everything about Spatial Analyst to get immediate results. Begin learning Spatial Analyst by reading Chapter 2, 'Quick-start tutorial'. This chapter introduces you to some of the tasks you can accomplish using Spatial Analyst and provides an excellent starting point as you start to think about how to tackle your own spatial problems. Spatial Analyst comes with the data used in the tutorial, so you can follow along step by step at your computer.

If you prefer to jump right in and experiment on your own, use Chapter 7, 'Performing spatial analysis', as a guide to learn the concepts and the steps to perform a certain task.

Finding answers to questions

Like most people, your goal is to complete your tasks while investing a minimum amount of time and effort on learning how to use software. You want intuitive, easy-to-use software that gives you immediate results without having to read pages of documentation. However, when you do have a question, you want the answer quickly so you can complete your task. That's what this book is all about—getting the answers you need, when you need them.

This book describes spatial analysis tasks—from basic to advanced—that you'll perform using Spatial Analyst. Although you can read this book from start to finish, you'll likely use it more as a reference. When you want to know how to perform a particular task, such as finding the shortest path, just look it up in the table of contents or the index. What you'll find is a concise, step-by-step description of how to complete the task. Some chapters also include detailed information that you can read if you want to learn more about the concepts behind the tasks. You may also refer to the glossary in this book if you come across any unfamiliar GIS terms or need to refresh your memory.

About this book

This book is designed to help you perform spatial analysis by giving you conceptual information and teaching you how to perform tasks to solve your spatial problems. Topics covered in Chapter 2, 'Quick-start tutorial', assume you are familiar with the fundamentals of GIS and have a basic knowledge of ArcGIS. If you are new to GIS or ArcMap™, you are encouraged to take some time to read *Getting Started with ArcGIS* and *Using ArcMap*, which you received in your ArcGIS package. It is not necessary to do so to continue with this book; simply use these books as references.

Chapter 3, 'Modeling spatial problems', takes you through the spatial modeling process, helping you to break down your spatial problems into manageable pieces. Chapter 4, 'Understanding raster data', helps you to understand raster data, and Chapter 5, 'Understanding cell-based modeling', explains the process of cell-based modeling. Chapter 6, 'Setting up your analysis environment', tells you how to set up your analysis options before performing analysis, and Chapter 7, 'Performing spatial analysis', provides detailed information to help you perform each spatial function.

The appendices are split into three sections: Appendix A explains Map Algebra syntax and rules for the Raster Calculator, Appendix B provides a table of supported operators and precedence values for use in the Raster Calculator, and Appendix C explains remap tables for use when reclassifying data using the Raster Calculator.

Getting help on your computer

In addition to this book, use the ArcGIS Desktop Help system to learn how to use Spatial Analyst and ArcMap. To learn how to use the ArcGIS Desktop Help system, see *Using ArcMap*.

Contacting ESRI

If you need to contact ESRI for technical support, see the product registration and support card you received with ArcGIS Spatial Analyst, or refer to 'Contacting Technical Support' in the 'Getting more help' section of the ArcGIS Desktop Help system. You can also visit ESRI on the Web at *www.esri.com* and *arconline.esri.com* for more information on Spatial Analyst and ArcGIS.

ESRI education solutions

ESRI provides educational opportunities related to geographic information science, GIS applications, and technology. You can choose among instructor-led courses, Web-based courses, and self-study workbooks to find education solutions that fit your learning style and pocketbook. For more information, go to *www.esri.com/education*.

Quick-start tutorial

2

IN THIS CHAPTER

- **Exercise 1: Displaying and exploring your data**
- **Exercise 2: Finding a site for a new school**
- **Exercise 3: Finding an alternative route to the new school site**

With Spatial Analyst you can easily perform spatial analysis on your data. You can provide answers to simple spatial questions, such as "How steep is it at this location?" or "What direction is this location facing?", or you can find answers to more complex spatial questions, such as "Where is the best location for a new facility?" or "What is the least-cost path from A to B?" When used in conjunction with ArcMap, Spatial Analyst provides a comprehensive set of tools for exploring and analyzing your spatial data, enabling you to find solutions to your spatial problems.

Tutorial scenario

The town of Stowe, Vermont, USA, has experienced a substantial increase in population. Demographic data suggests this increase has occurred due to families with children moving to the region, taking advantage of the many recreational facilities located nearby. It has been decided that a new school must be built to take the strain off the existing schools, and as a town planner you have been assigned the task of finding the potential sites.

Spatial Analyst provides the tools to find an answer to such spatial problems. This tutorial will show you how to use some of these tools and will give you a solid basis from which you can start to think about how to solve your own specific spatial problems.

It is assumed that you have installed the Spatial Analyst extension before you begin this tutorial. The data required is included on the Spatial Analyst installation disk (the default installation path is ArcGIS\ArcTutor\Spatial, on the drive where the tutorial data is installed). The datasets were provided courtesy of the State of Vermont for use in this tutorial. The tutorial scenario is fictitious, and the original data has been adapted for the purpose of the tutorial.

The datasets are:

Dataset	Description
Elevation	Raster dataset of the elevation of the area
Landuse	Raster dataset of the landuse types over the area
Roads	Feature dataset displaying linear road network
Rec_sites	Feature dataset displaying point locations of recreation sites
Schools	Feature dataset displaying point locations of existing schools
Destination	Feature dataset displaying the destination point for use in finding the shortest path

In this tutorial you will first explore your data to learn more about it and to understand its relationships. Then, you will find suitable locations for the new school, based on the fact that it is preferable to locate close to recreational facilities for ease of access to these places for the children, and it is also important to locate away from existing schools to spread their locations over the town. You also want to avoid steep slopes and certain landuse types.

Once you have found the best sites, you will examine these locations to see which is potentially the most suitable. You will then examine the data to see if any problems may arise from building the school in the chosen location.

This tutorial is divided into exercises and is designed to let you explore the functionality of Spatial Analyst at your own pace.

- Exercise 1 shows you how to display and explore your data using the functionality of ArcMap and Spatial Analyst. You will add and display your datasets, highlight values on the map, identify locations to obtain values, examine a histogram, and create a hillshade.
- Exercise 2 helps you to find the best location for a new school by creating a suitability map. You will derive datasets of distance and slope, reclassify datasets to a common scale, weight those that are more important to consider, then combine the datasets to find the most suitable locations.
- Exercise 3 shows you how to find an alternative route—the least-cost, or shortest path—for a road to the new school site.

Copies of the results obtained from each exercise are stored in the Results folder on your local drive where you installed the tutorial data (the default installation path is ArcGIS\ArcTutor\Spatial\Results).

You will need about one hour of focused time to complete the tutorial. However, you can perform the exercises one at a time if you wish, saving your results along the way when recommended.

Exercise 1: Displaying and exploring your data

You should explore your data to understand it and to identify relationships. Understanding your data and recognizing relationships will enable you to more accurately prepare your data for analysis.

In this exercise, you will open ArcMap and add the Spatial Analyst toolbar to your ArcMap session. You will then explore your datasets using functionality within ArcMap and Spatial Analyst.

Starting ArcMap and Spatial Analyst

1. Start ArcMap by either double-clicking a shortcut installed on your desktop or using the Programs list in your Start menu.
2. Click OK to open a new empty map.

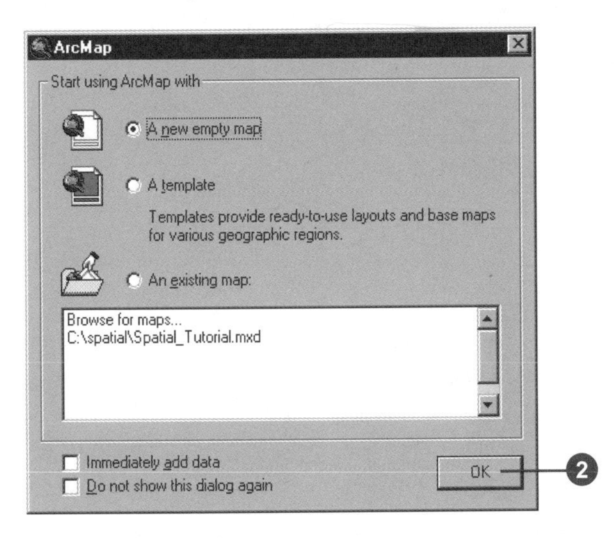

3. Click View, point to Toolbars, and click Spatial Analyst.

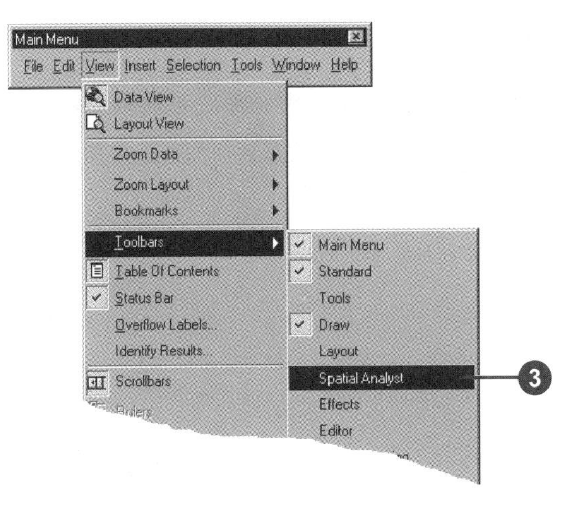

The Spatial Analyst toolbar is added to your ArcMap session.

Enabling the Spatial Analyst toolbar

1. Click the Tools menu.
2. Click Extensions and check Spatial Analyst.
3. Click Close.

Adding data to your ArcMap session

1. Click the Add Data button on the Standard toolbar.

2. Navigate to the folder on your local drive where you installed the tutorial data (the default installation path is ArcGIS\ArcTutor\Spatial, on the drive where the tutorial data is installed).
3. Click elevation, press and hold down the Shift key, then click landuse, rec_sites, roads, and schools.
4. Click Add.

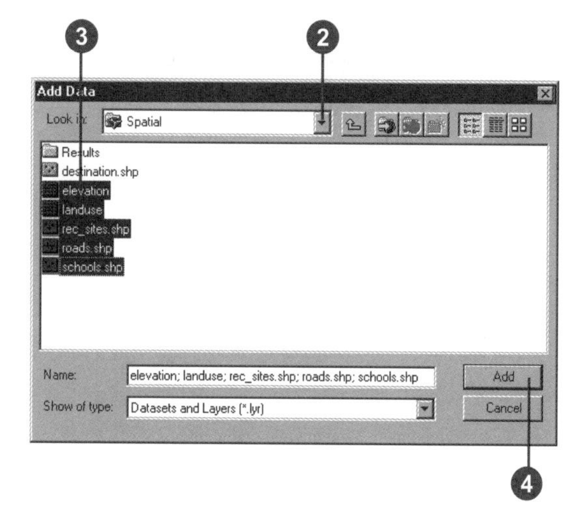

The datasets are added to the ArcMap *table of contents* as layers.

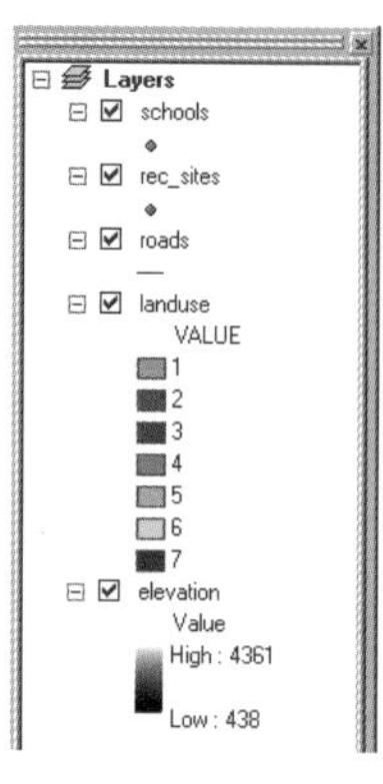

Displaying and exploring data

You will now explore the display capabilities of ArcMap by changing the *symbology* of some of the layers.

1. Right-click landuse in the table of contents and click Properties.

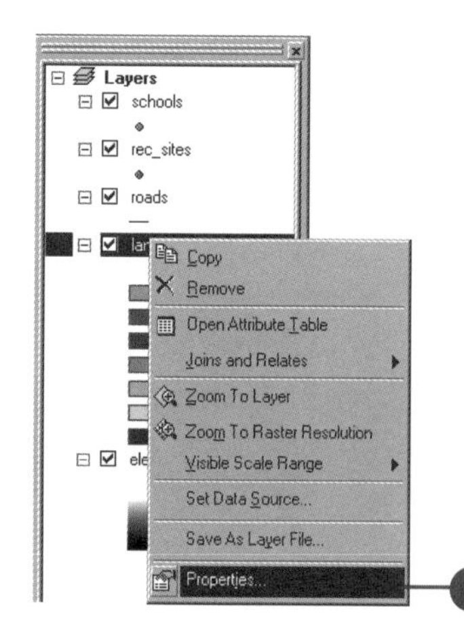

2. Click the Symbology tab.

   All landuse categories are currently drawn using *cell* values as the Value Field and in random colors. You will change the Value Field to be more meaningful and change the color of each *symbol* to show a more appropriate color for each landuse on the map.
3. Click the Value Field dropdown arrow and click landuse.
4. Double-click each symbol and choose a suitable color to represent each landuse type.
5. Click OK.

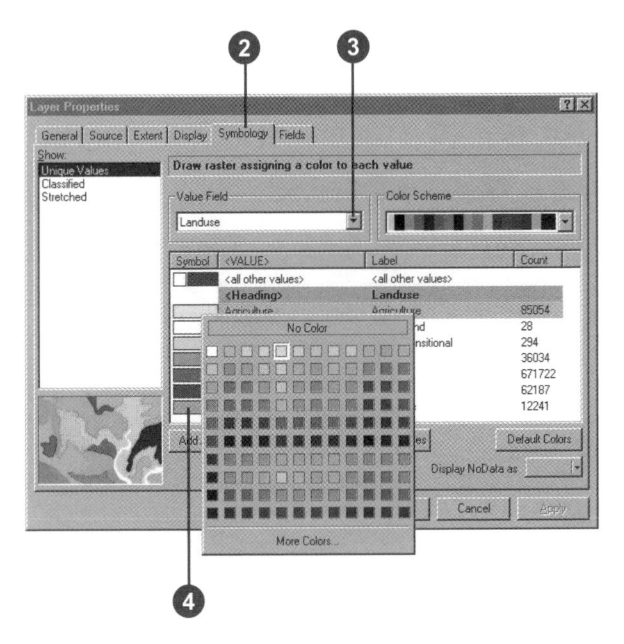

The changes you make are reflected in the table of contents and in the map.

You can also change the color and properties of symbols via the table of contents.

6. Click the point representing schools in the table of contents.

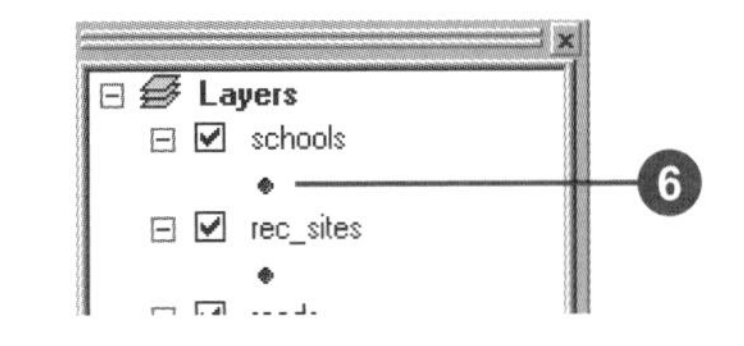

7. Scroll to the School 2 symbol and click it.
8. Click the color dropdown arrow and click a color.
9. Click OK.

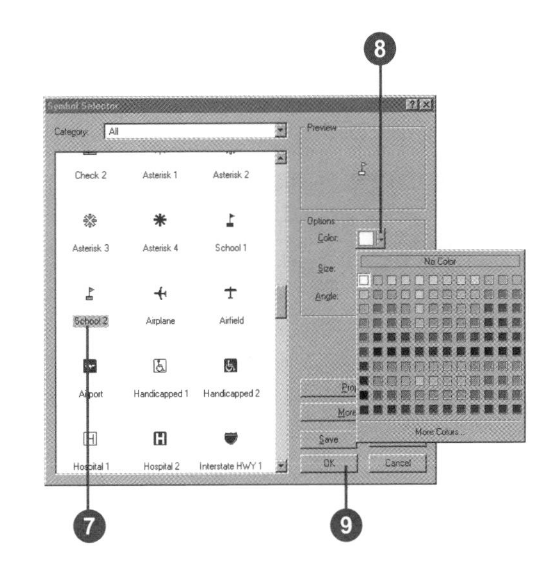

The changes you make are reflected in the table of contents and in the map.

Highlighting a selection on the map

Examining the attribute table gives you an idea of the number of cells of each attribute in the dataset.

1. Right-click landuse in the table of contents and click Open Attribute Table.

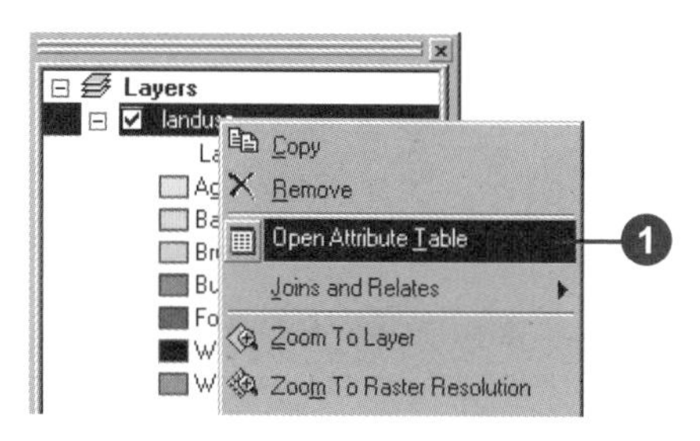

Notice that Forest (value of 6) has the largest count, followed by Agriculture (value of 5), then Water (value of 2).

Attributes of landuse

ObjectID*	Value	Count	Landuse
1	1	294	Brush/transitional
2	2	62187	Water
3	3	28	Barren land
4	4	36034	Built up
5	5	85054	Agriculture
6	6	671722	Forest
7	7	12241	Wetlands

Record: 1 Show: All Selected Records (0 out of 7 Selected.) Options

2. Click the row representing Wetlands (value of 7).

Attributes of landuse

ObjectID*	Value	Count	Landuse
1	1	294	Brush/transitional
2	2	62187	Water
3	3	28	Barren land
4	4	36034	Built up
5	5	85054	Agriculture
6	6	671722	Forest
7	7	12241	Wetlands

Record: 1 Show: All Selected Records

2

This *selected set,* all areas of Wetlands, is highlighted on the map.

3. Click the Options button on the Open Attribute Table dialog box, then click Clear Selection.

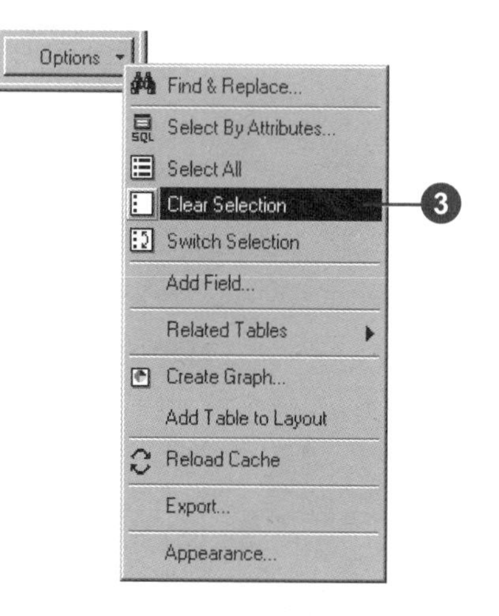

4. Click the Close button to close the Attributes of landuse table.

Identifying features on the map

1. Click the Identify tool on the Tools toolbar.
2. Click the Rec_site shown in the map below to identify the features in this particular location.

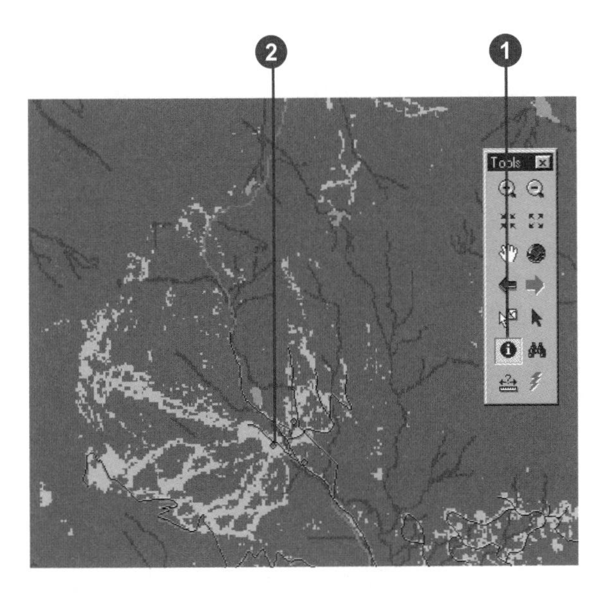

Note: Your display will not be zoomed in this much; this is only to show the location of the recreation site to click.

3. Click the Layers dropdown arrow on the Identify Results dialog box and click All layers.
4. Click the Rec_site again to identify the features in this particular location for all layers.
5. Expand the tree of each layer to obtain the value for each layer in this location.
6. Close the Identify Results dialog box.

Field	Value
FID_1	20
Shape	Point
FID	315
AREA	0
PERIMETER	0
MAP	90021
SITE_NAME	Stowe Mountain Resort
FIPS_CODE	15040
ALT_NAME	
S_ADDRESS	5781 Mountain Road

Using Spatial Analyst to explore your data

You will now create a *histogram* from the landuse layer and a *hillshade* from the Elevation layer to gain more of an understanding of the nature of the landscape.

Setting the analysis properties

Before you use Spatial Analyst, you should set up the analysis options, stating the working directory, the extent, and the *cell size* for your analysis results. These settings are specified in the Options dialog box.

1. Click the Spatial Analyst dropdown arrow and click Options.

2. Specify a working directory on your local drive in which to place your analysis results. For example, type c:\spatial to create a folder called spatial on your C:\ drive for use throughout this tutorial.

3. Click the Extent tab.

4. Click the Analysis extent dropdown arrow and click Same as Layer “landuse”.

   The extent of all subsequent resulting datasets will be the same as the landuse layer.

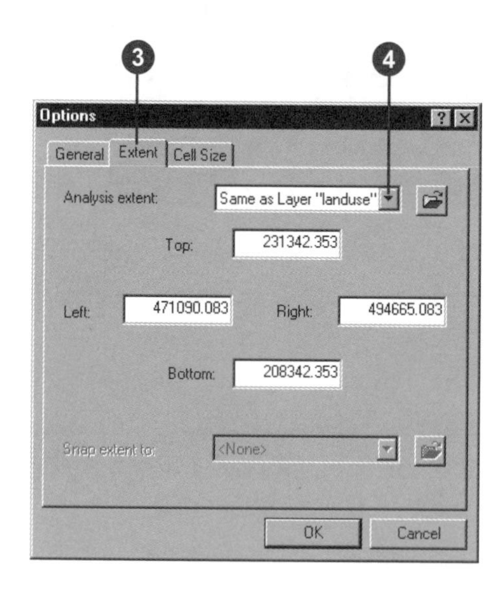

5. Click the Cell Size tab.
6. Click the Analysis Cell Size dropdown arrow and click Same as Layer “elevation”.
7. Click OK on the Options dialog box.

   This will set the cell size for your analysis results to be at a 30-meter resolution, the largest cell size of your datasets.

Examining a histogram

1. Click the Layer dropdown arrow and click landuse.
2. Click the Histogram button.

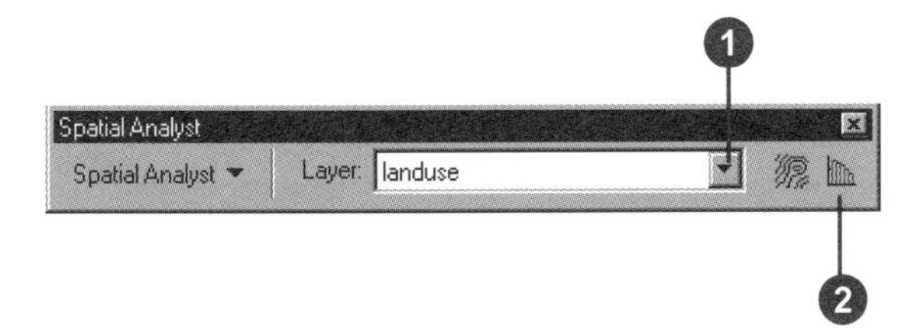

   The histogram displays the number of cells of each type of landuse.
3. Close the Histogram.

Creating a hillshade

Creating a hillshade from elevation data and adding transparency gives you a good visual impression of the terrain and can greatly enhance the display of your map.

1. Click the Spatial Analyst dropdown arrow, point to Surface Analysis, and click Hillshade.

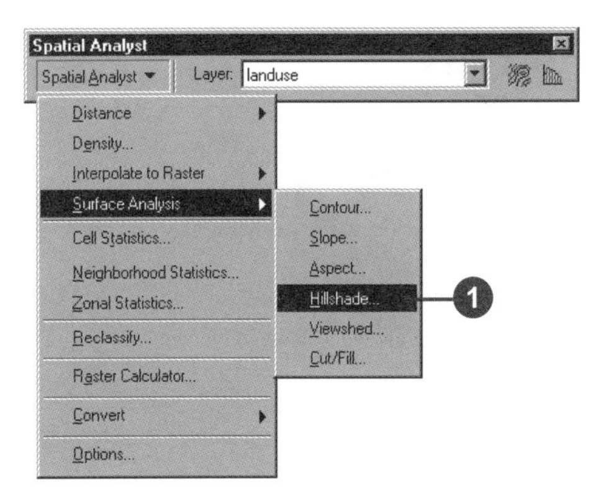

2. Click the Input surface dropdown arrow and click elevation. Leave the defaults for all other options.

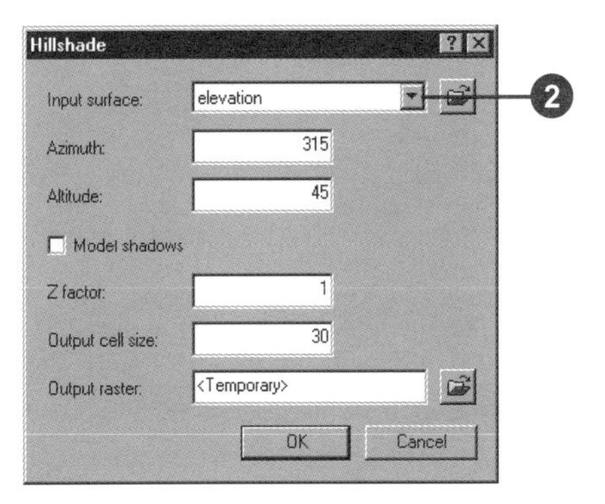

3. Click OK on the Hillshade dialog box.

   The result of the Hillshade function is added to the map as a new layer.

   All results from analysis functions are *temporary*. If you want to make any result available for future use, you should make the dataset *permanent*.

4. Right-click the created hillshade layer and click Make Permanent.
5. Navigate to the folder on your local drive where you set up your working directory (C:\Spatial).
6. Type "Hillshade" in the Name text box.
7. Click the Save as type dropdown arrow and click ESRI GRID.
8. Click Save.

   Note: A copy of Hillshade can be found in the location ArcGIS\ArcTutor\Spatial\Results\Ex1\Hillshade on the drive where the tutorial data is installed.

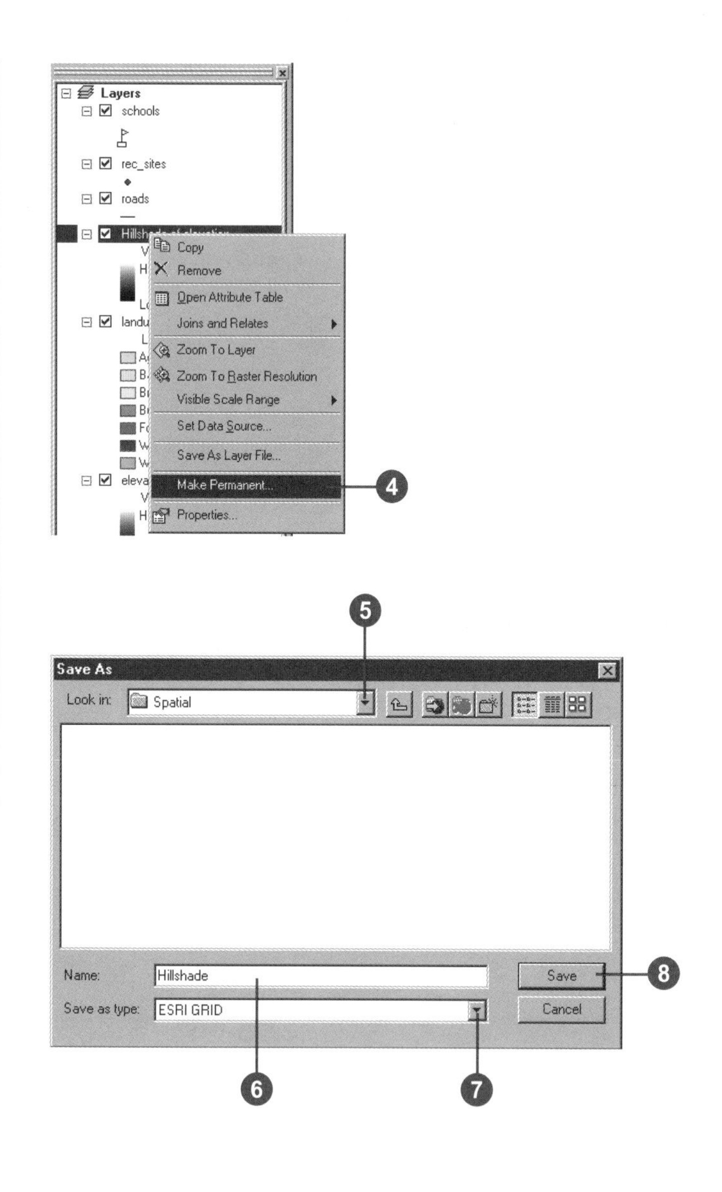

Applying transparency

You will now make the landuse layer transparent so the Hillshade can be seen through it.

1. Click Hillshade of elevation in the table of contents and drag the layer below the landuse layer.

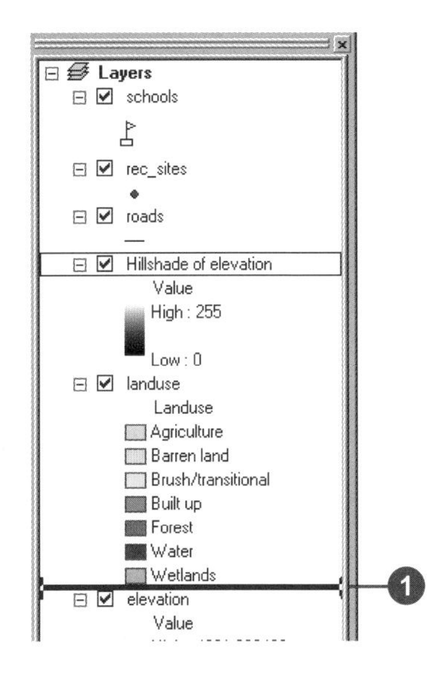

2. Click View on the Main menu, point to Toolbars, and click Effects.
3. Click the Layer dropdown arrow and click landuse.
4. Click the Adjust Transparency button and move the scroll bar up to 30 percent transparency.

   The Hillshade layer can now be seen underneath the landuse layer, giving a vivid impression of the terrain.

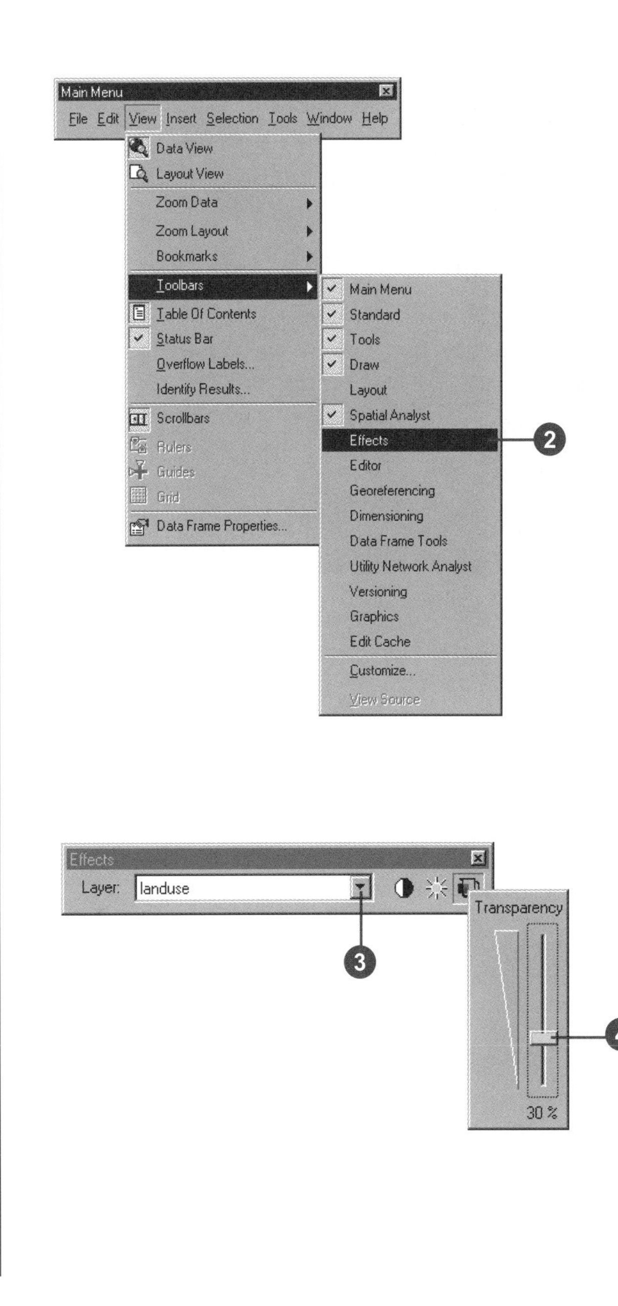

Exploring your data gives you a useful basis of information that will help you during your analysis. For example, you need to know the different landuse types and their distribution over an area, as well as their relative importance, in order to decide how much weight each should have in a *suitability model*. Alternatively, you need to know how rugged the terrain is so you know to include *slope* as a factor in determining the *least-cost path*.

Having explored your data, you are now in a position to begin to find suitable locations for the new school.

First, you will need to remove all the layers used in this exercise.

5. Click the top layer in the table of contents to highlight it. Press and hold the Shift key and click all other layers.
6. Right-click one of the layers in the table of contents and click Remove.

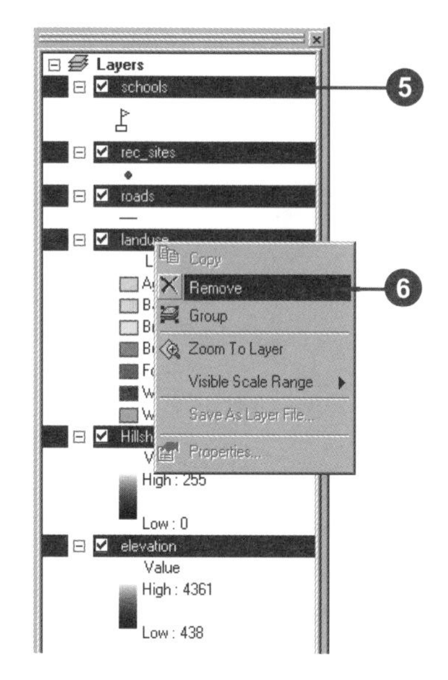

All layers will be removed from the ArcMap *data frame*.

This exercise showed you how to display and explore your data. In the next exercise you will use the Spatial Analyst functions to find a potential site for a new school. You can continue on with the tutorial or close ArcMap and continue at a later date. There is no need to save the map document at this point.

Note: To save your work at any time, click the File menu and click Save As. Navigate to the location where you set up your local working directory (C:\Spatial), specify a filename for the *map document*—Spatial_Tutorial—and click Save. Simply open Spatial_Tutorial.mxd when you wish to continue with the tutorial. You will, however, be prompted when it is appropriate to save the map document.

Exercise 2: Finding a site for a new school in Stowe, Vermont, USA

In this exercise you will find suitable locations for a new school. The four steps to produce such a suitability map are outlined below.

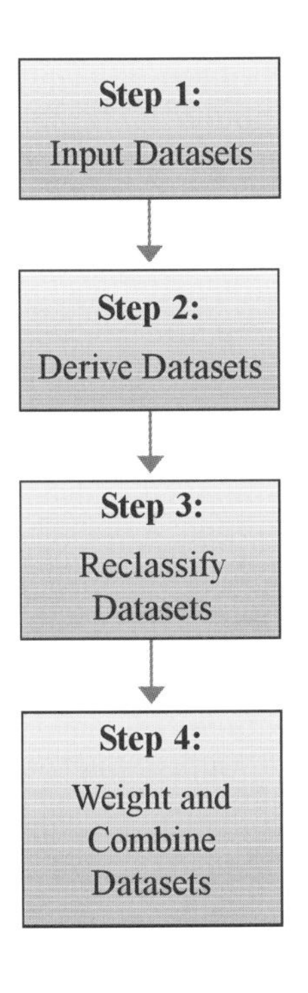

Decide which datasets you need as inputs. The datasets you will use in this exercise are displayed to the right.

Derive datasets. Create data from existing data to gain new information.

Reclassify each dataset to a common scale—for example, 1–10—giving higher values to more suitable attributes.

Weight datasets that should have more influence in the suitability model if necessary, then combine them to find the suitable locations.

Your input datasets in this exercise are Landuse, Elevation, Recreation Sites, and Existing Schools. You will derive slope, distance to recreation sites, and distance to existing schools, then *reclassify* these derived datasets to a common scale from 1–10. You will then weight them according to a percentage influence and combine them to produce a map displaying suitable locations for the new school. The diagram to the right shows the process you will take.

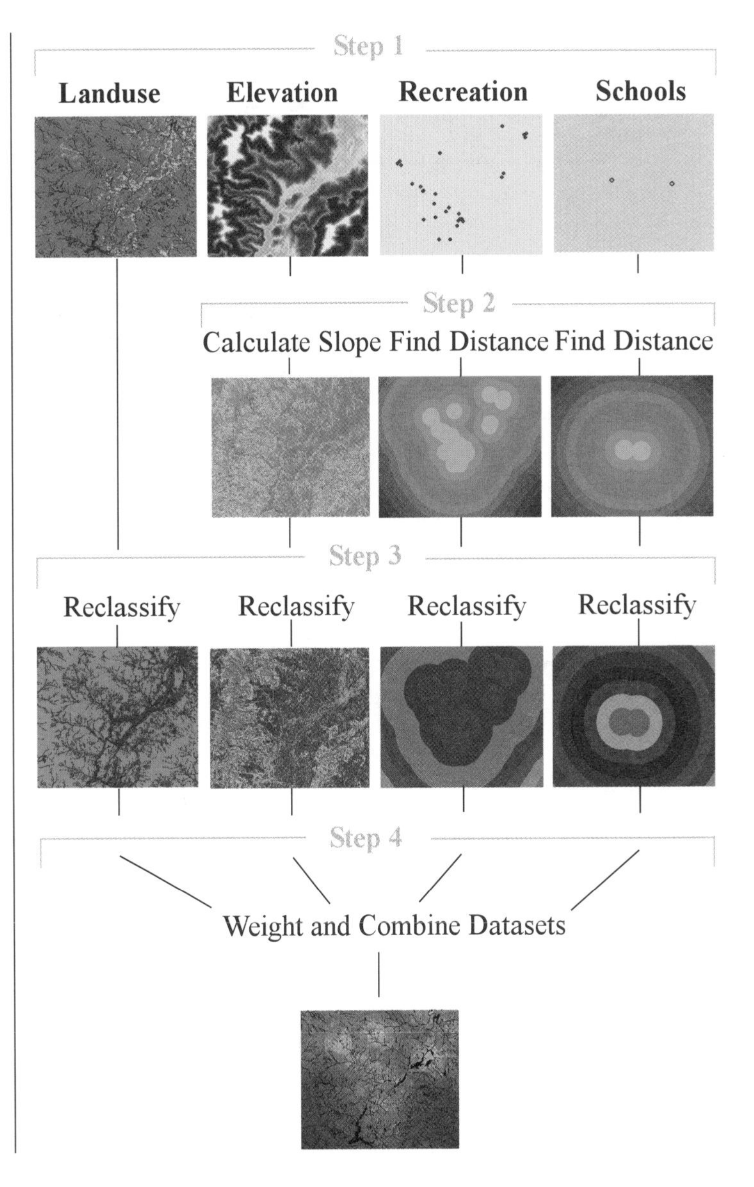

Step 1: Inputting datasets

1. Click the Add Data button on the Standard toolbar.

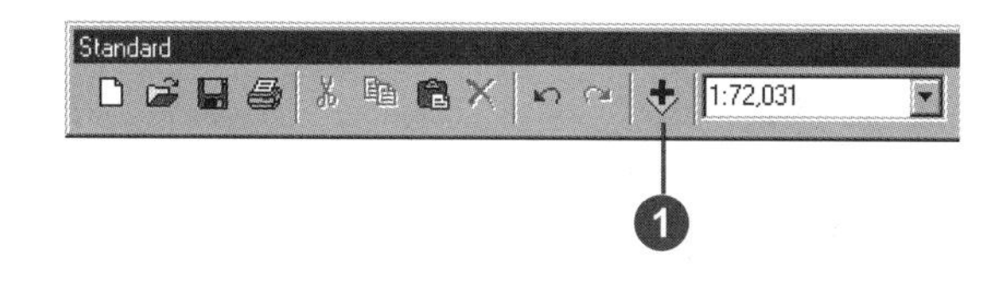

2. Navigate to the folder on your local drive where you installed the tutorial data (the default installation path is ArcGIS\ArcTutor\Spatial, on the drive where the tutorial data is installed).
3. Click elevation, then click and hold down the Ctrl key and click landuse, rec_sites, and schools.
4. Click Add.

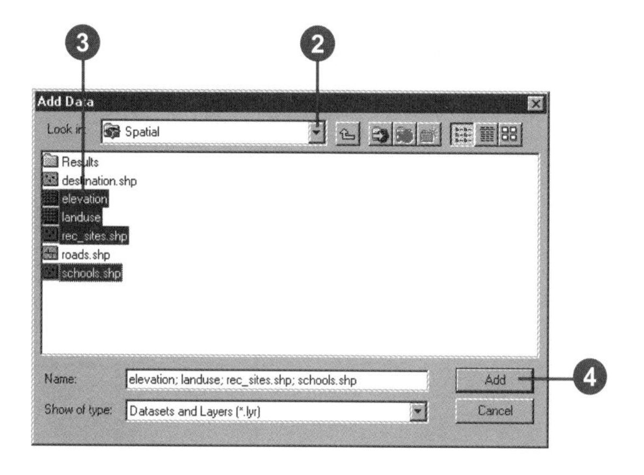

Each dataset is added to the ArcMap table of contents as a *layer*.

Setting the analysis properties

Set up the analysis options like you did in Exercise 1.

1. Click the Spatial Analyst dropdown arrow and click Options.
2. Specify a working directory on your local drive in which to place your analysis results. Type c:\spatial to create a folder called spatial on your C:\ drive.
3. Click the Extent tab.
4. Click the Analysis Extent dropdown arrow and click Same as Layer "landuse".
5. Click the Cell Size tab.
6. Click the Analysis Cell Size dropdown arrow and click Same as Layer "elevation".
7. Click OK on the Options dialog box.

Step 2: Deriving datasets

Deriving data from your input datasets is the next step in the suitability model. You will derive the following:

- Slope from elevation
- Distance from recreation sites
- Distance from existing schools

Deriving slope

Since the area is mountainous, you need to find areas of relatively flat land to build on, so you will take into consideration the slope of the land.

1. Click the Spatial Analyst dropdown arrow, point to Surface Analysis, and click Slope.

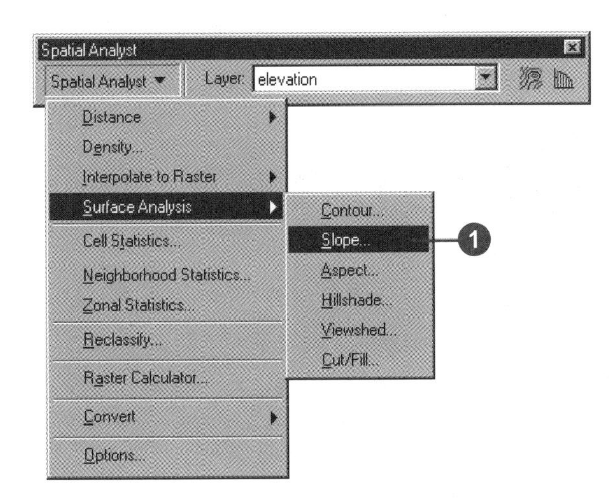

2. Click the Input surface dropdown arrow and click elevation.

3. Type slope in the Output raster text box to permanently save your output slope dataset to the location of your working directory (c:\spatial).

You will use this dataset again in Exercise 3.

Note: A copy of this slope dataset can be found in the location ArcGIS\ArcTutor\Spatial\Results\Ex2\Slope.

4. Click OK.

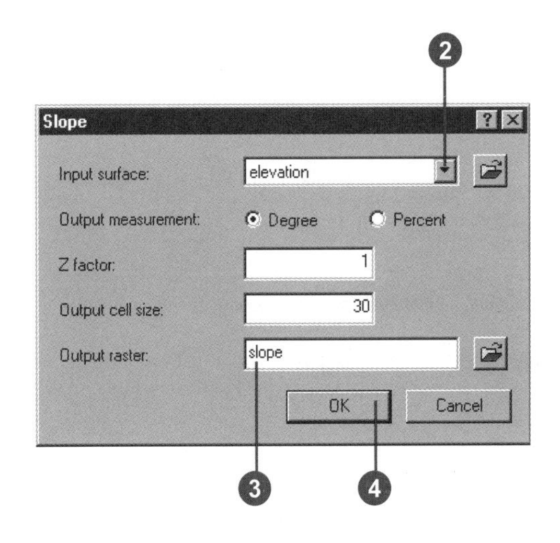

The output slope dataset will be added to your ArcMap session as a new layer. High values—red areas—indicate steeper slopes.

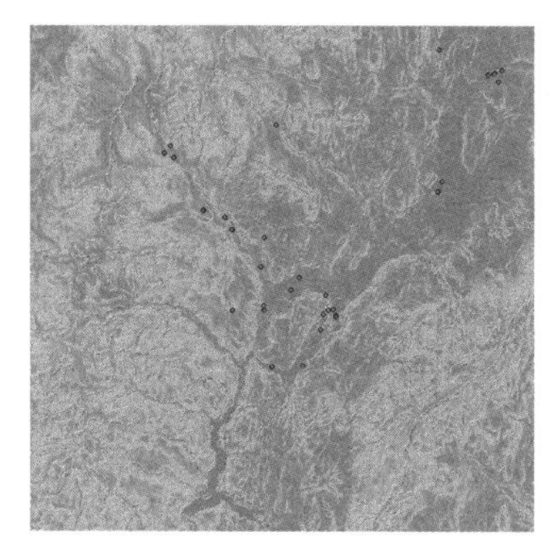

Deriving distance from recreation sites

In this model, it is preferable that the school be built near recreational facilities, so you will now calculate the straight-line distance from Recreation Sites.

1. Click the Spatial Analyst dropdown arrow, point to Distance, and click Straight Line.

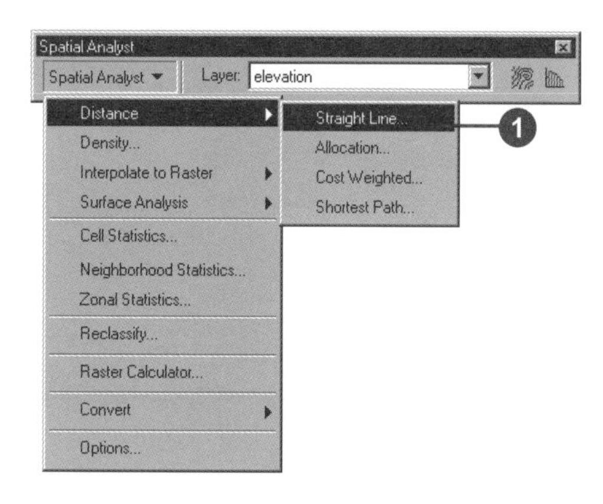

2. Click the Distance to dropdown arrow and click rec_sites.

   Leave the defaults for all other options.

3. Click OK.

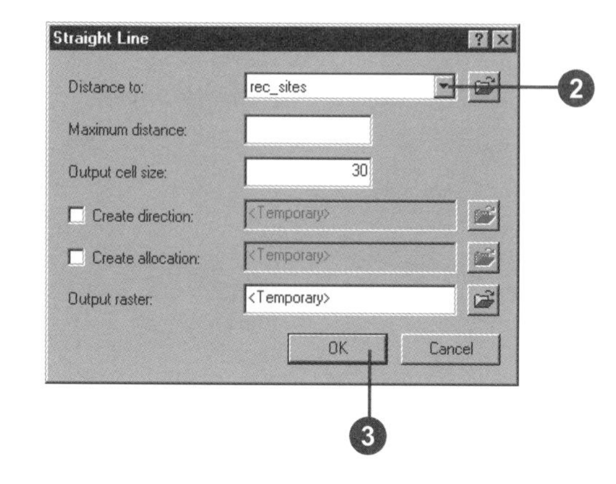

The output distance to the rec_sites dataset will be added to your ArcMap session as a new layer. Values of zero indicate the location of a recreation site, with values—distances—increasing as you move away from each of these sites.

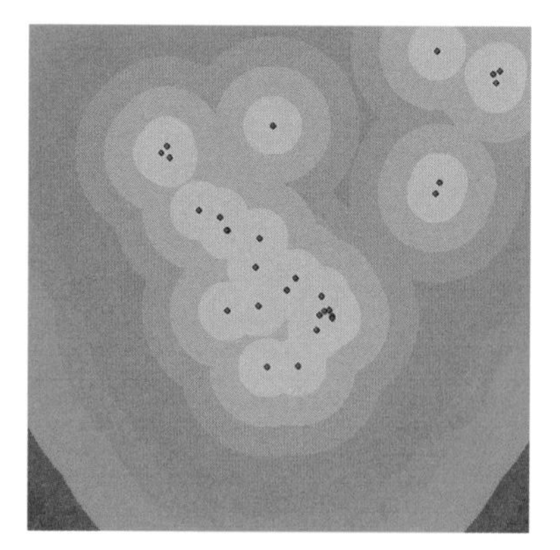

Note: A copy of this distance to rec_sites dataset can be found in the location ArcGIS\ArcTutor\Spatial\Results\Ex2\recD.

4. Uncheck the box next to Schools to turn off this layer so you only see the locations of the recreation sites and the distance to them.

Deriving distance from schools

You will now derive a dataset of distance from existing schools. It is preferable to locate the new school away from existing schools to spread out their locations through the town.

1. Click the Spatial Analyst dropdown arrow, point to Distance, and click Straight Line.
2. Click the Distance to dropdown arrow and click schools.

   Leave the defaults for all other options.
3. Click OK.

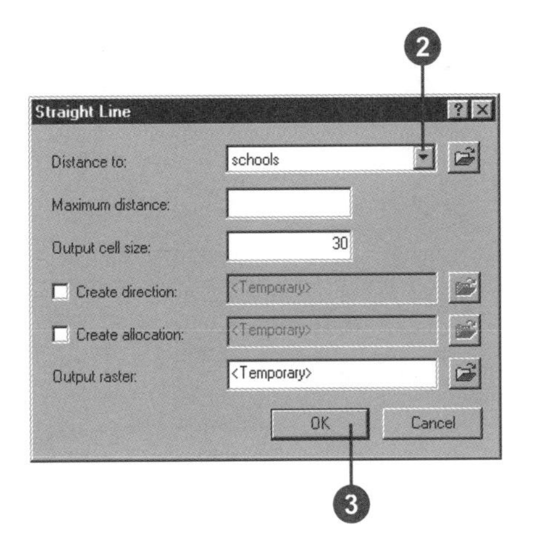

The output distance to schools dataset will be added to your ArcMap session as a new layer.

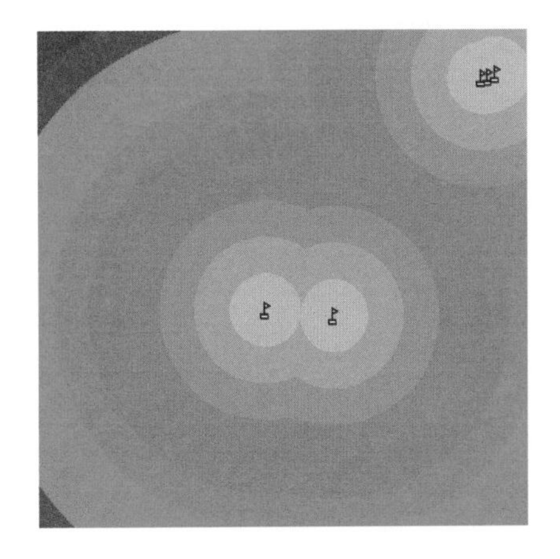

4. Check the box next to the schools layer to turn it back on and uncheck the box next to rec_sites to turn this layer off so you only see the locations of the schools and the distance to them.

Note: A copy of this distance to schools dataset can be found in the location ArcGIS\ArcTutor\Spatial\Results\Ex2\schD.

Step 3: Reclassifying datasets

You now have the required datasets to find the best location for the new school. The next step is to combine them to find out where the potential locations can be found.

In order to combine the datasets, they must first be set to a common scale. That common scale is how suitable a particular location—each cell—is for building a new school. You will reclassify each dataset to a common scale, within the range 1–10, giving higher values to attributes within each dataset that are more suitable for locating the school:

- Reclassify slope
- Reclassify Distance to recreation sites
- Reclassify Distance to schools
- Reclassify landuse

Reclassifying slope

It is preferable that the new school site be located on relatively flat ground. You will reclassify the Slope layer, giving a value of 10 to the most suitable slopes—those with the lowest angle of slope—and 1 to the least suitable slopes—those with the steepest angle of slope.

1. Click the Spatial Analyst dropdown arrow and click Reclassify.
2. Click the Input raster dropdown arrow and click Slope.
3. Click Classify.

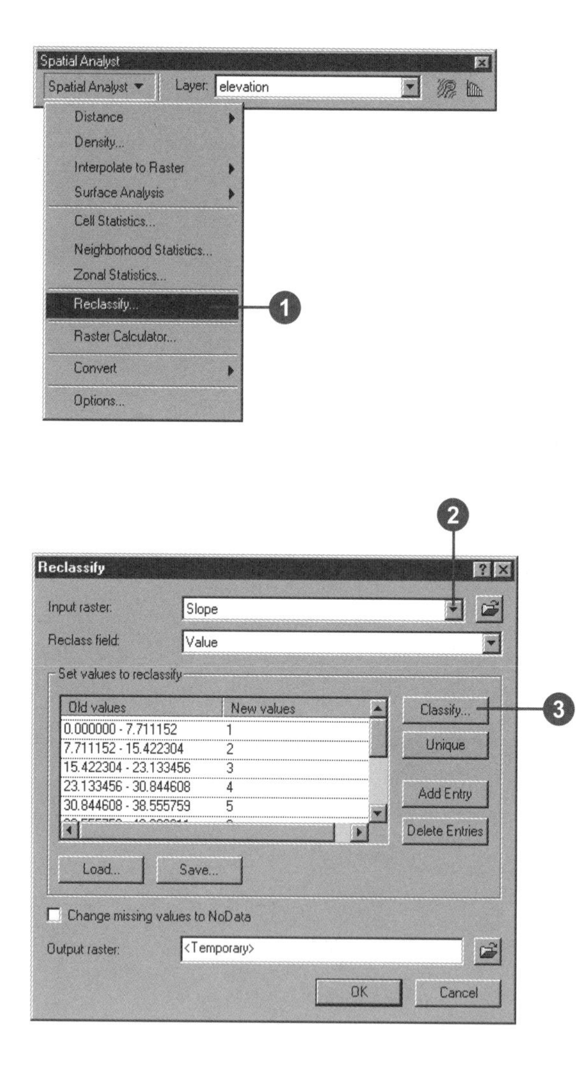

4. Click the Method dropdown arrow and click Equal Interval.
5. Click the Classes dropdown arrow and click 10.
6. Click OK.

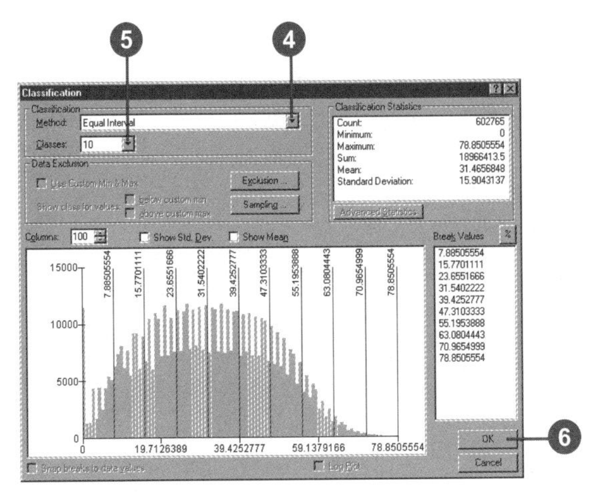

You want to reclassify the Slope layer so steep slopes are given low values, as these are least suitable for building on.

7. Click the first New value record in the Reclassify dialog box and change it to a value of 10. Give a value of 9 to the next New value, 8 to the next, and so on. Leave NoData as NoData.
8. Click OK.

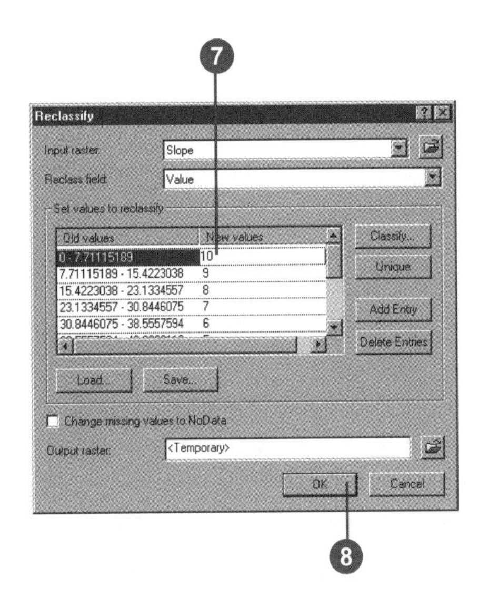

The output reclassified slope dataset will be added to your ArcMap session as a new layer. Locations with higher values—less-steep slopes—are more suitable than locations with lower values—steeper slopes.

Note: A copy of this reclassified slope dataset can be found in the location ArcGIS\ArcTutor\Spatial\Results\Ex2\slopeR.

Reclassifying distance to recreation sites

The school should be located near recreational facilities. You will reclassify this dataset, giving a value of 10 to areas closest to recreation sites—the most suitable locations—giving a value of 1 to areas far from recreation sites—the least suitable locations—and ranking the values in between. By doing this you will find out which areas are near and which areas are far from recreation sites.

1. Click the Spatial Analyst dropdown arrow and click Reclassify.

2. Click the Input raster dropdown arrow and click Distance to rec_sites.
3. Click Classify.

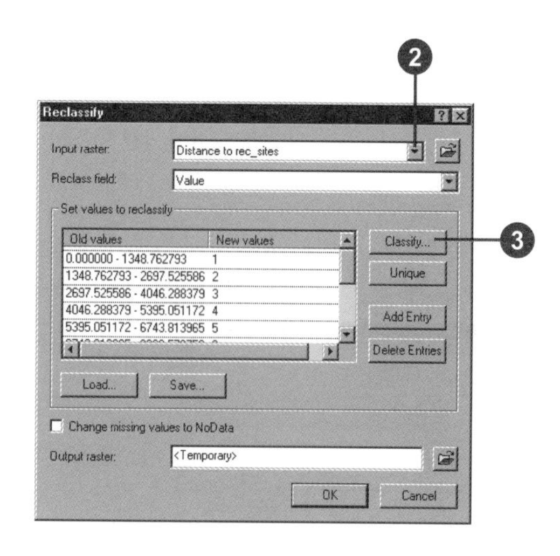

4. Click the Method dropdown arrow and click Equal Interval.
5. Click the Classes dropdown arrow and click 10.
6. Click OK.

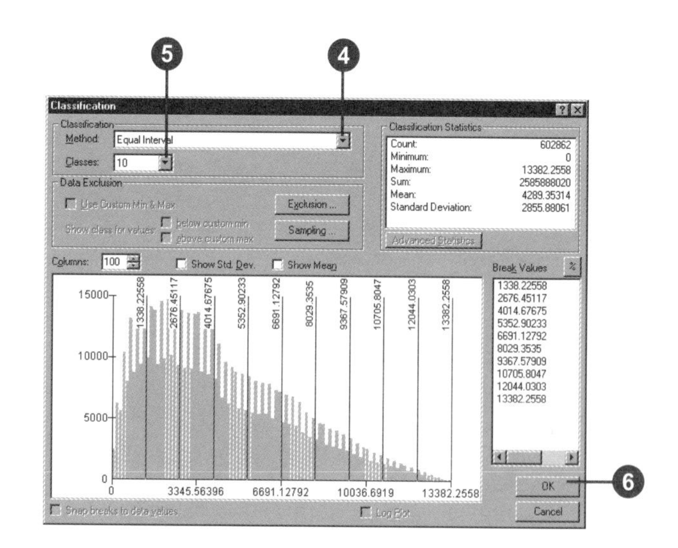

You want to locate the school near recreational facilities, so you will give higher values to locations close to recreational facilities, as these are the most desirable.

7. As you did when reclassifying the Slope layer, click the first New value record in the dialog box and change it to a value of 10. Give a value of 9 to the next New value, 8 to the next, and so on. Leave NoData as NoData.
8. Click OK.

The output reclassified distance to recreation sites dataset will be added to your ArcMap session as a new layer. It shows locations that are more suitable for locating another school. High values indicate more suitable locations.

Note: A copy of this reclassified distance from recreation sites dataset can be found in the location ArcGIS\ArcTutor\Spatial\Results\Ex2\recR.

Reclassifying distance to schools

It is necessary to locate the new school away from existing schools in order to avoid encroaching on their catchment areas. You will reclassify the Distance to schools layer, giving a value of 10 to areas away from existing schools—the most suitable locations—giving a value of 1 to areas near existing schools—east suitable locations—and ranking the values in between. By doing this you will find out which areas are near and which areas are far from existing schools.

1. Click the Spatial Analyst dropdown arrow and click Reclassify.

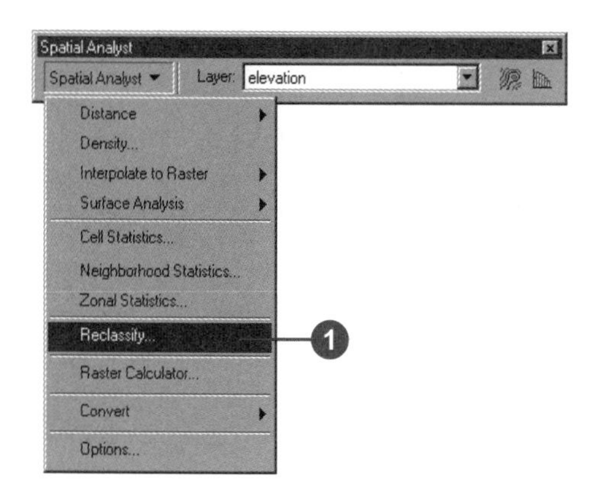

2. Click the Input raster dropdown arrow and click Distance to schools.
3. Click Classify.

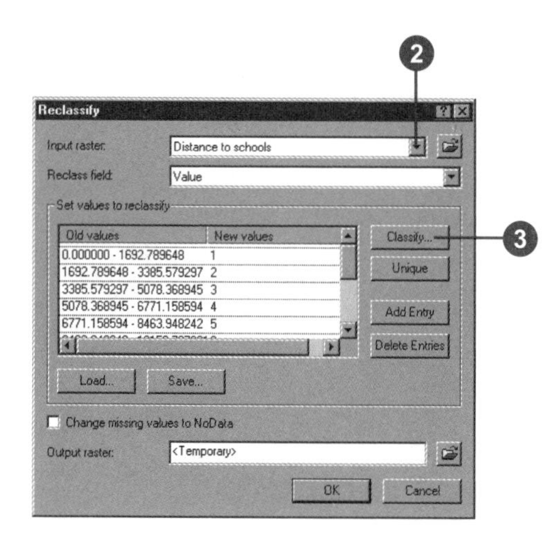

4. Click the Method dropdown arrow and click Equal Interval.
5. Click the Classes dropdown arrow and click 10.
6. Click OK.

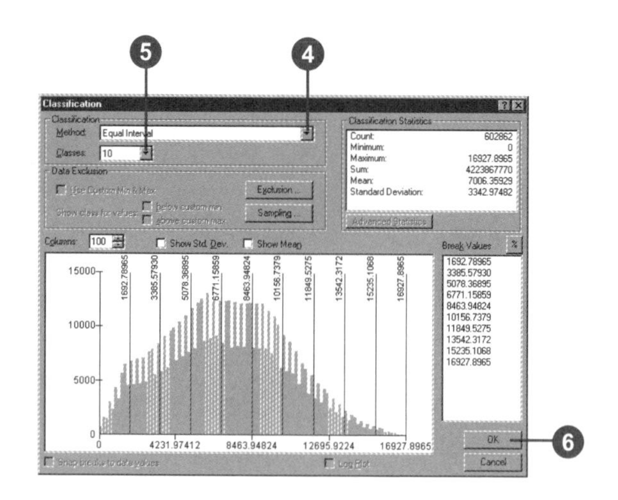

You want to locate the school away from existing schools, so you will give higher values to locations farther away, as these locations are most desirable.

As the default gives high New values—more suitable—to high Old values—locations farther away from existing schools—you do not need to change any values this time.

7. Click OK.

The output reclassified distance to schools dataset will be added to your ArcMap session as a new layer. It shows locations that are more suitable for locating another school. Higher values indicate more suitable locations.

Note: A copy of this reclassified distance from schools dataset can be found in the location ArcGIS\ArcTutor\Spatial\Results\Ex2\schR.

Reclassifying landuse

At a town planners meeting it was decided that certain landuse types were better for building on than others, taking into consideration the costs involved in building on different landuse types.

You will now reclassify landuse. A lower value indicates that a particular landuse type is less suitable for building on. Water and Wetlands will be given NoData as they cannot be built on and should be excluded.

1. Click the Spatial Analyst dropdown arrow and click Reclassify.

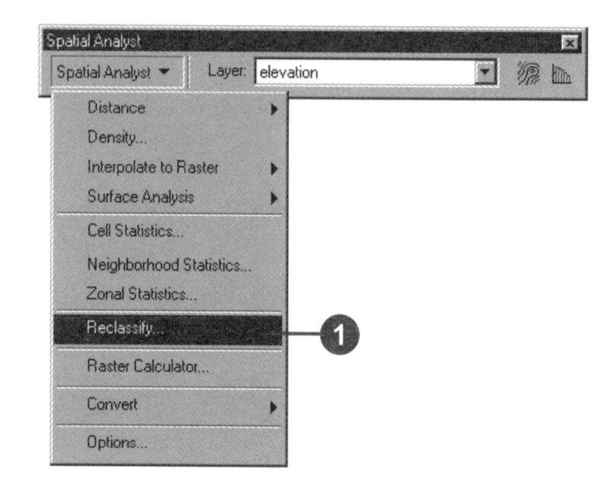

2. Click the Input raster dropdown arrow and click landuse.
3. Click the Reclass field dropdown arrow and click Landuse.
4. Type the following values in the New values column:

Agriculture—10 | Built up—3

Barren land—6 | Forest—4

Brush/Transitional—5

You will now remove the Water and Wetland attributes and change their values to NoData.

5. Click the row for Water, press the Shift key, click Wetlands, then click Delete Entries.
6. Check Change missing values to NoData.

   All values for Water and Wetlands will be changed to NoData.
7. Click OK.

The output reclassified landuse dataset will be added to your ArcMap session as a new layer. It shows locations that have landuse types that are considered to be better than others for locating the school—higher values indicate more suitable locations.

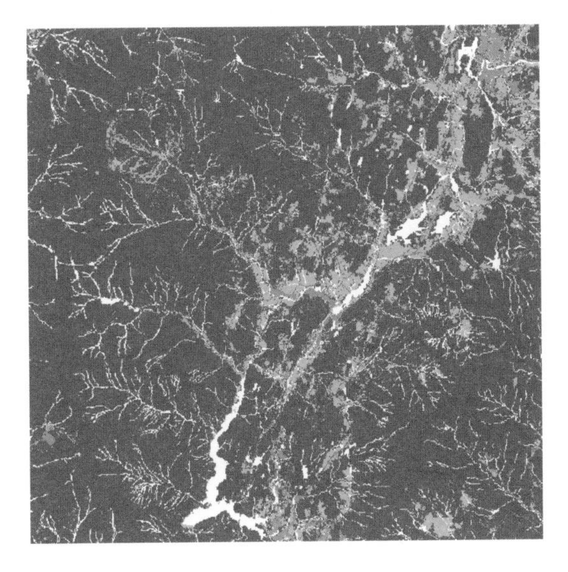

8. Right-click Reclass of landuse in the table of contents and click Properties.
9. Click the Symbology tab.
10. Click the Display NoData as dropdown arrow and click Arctic White to show NoData values—Water and Wetlands—in this color.
11. Click OK.

9

Layer Properties

General | Source | Extent | Display | Symbology | Fields

Show:
Unique Values
Classified
Stretched

Draw raster assigning a color to each value

Value Field: Value

Color Scheme

Symbol	<VALUE>	Label	Count
	<all other values>	<all other values>	
	<Heading>	VALUE	
	3	3	25054
	4	4	466816
	5	5	205
	6	6	21
	10	10	59224

Add All Values | Add Values... | Remove Values | Default Colors

Display NoData as

OK | Cancel | Apply

11 10

Note: A copy of this reclassified landuse dataset can be found in the location ArcGIS\ArcTutor\Spatial\Results\Ex2\landuseR.

Step 4: Weighting and combining datasets

After applying a common scale to your datasets, where higher values are given to those attributes that are considered more suitable within each dataset, you are ready to combine them to find the most suitable locations.

If all datasets were equally important, you could simply combine them at this point; however, you have been informed that it is preferable to locate the new school close to recreational facilities and away from other schools. You will weight all the datasets, giving each a percentage influence. The higher the percentage, the more influence a particular dataset will have in the suitability model.

You will give the layers the following percent influence:

(Each percentage is divided by 100 to *normalize* the values.)

Reclass of Distance to rec_sites:	0.5	(50%)
Reclass of Distance to schools:	0.25	(25%)
Reclass of landuse:	0.125	(12.5%)
Reclass of slope:	0.125	(12.5%)

1. Click the Spatial Analyst dropdown arrow and click *Raster Calculator*.

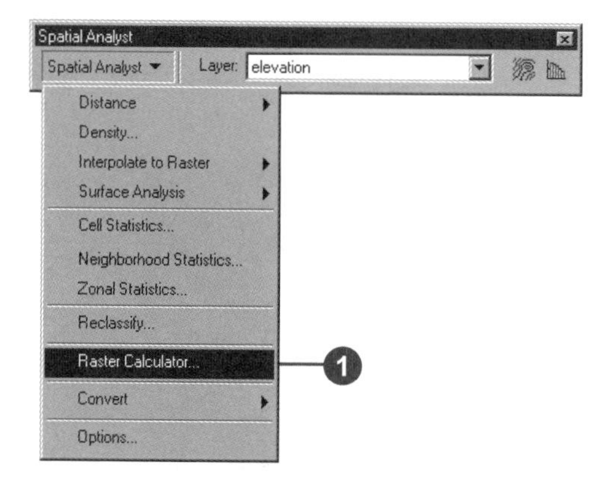

2. Double-click Reclass of Distance to rec_sites from the Layers list to add it to the expression box.
3. Click Multiply.
4. Click 0.5.
5. Click Add.
6. Double-click Reclass of Distance to schools.
7. Click Multiply.
8. Click 0.25.
9. Click Add.
10. Double-click Reclass of landuse.
11. Click Multiply.
12. Click 0.125.
13. Click Add.
14. Double-click Reclass of slope.
15. Click Multiply.

16. Click 0.125.
17. Click Evaluate to perform the weighting and combining of the datasets.

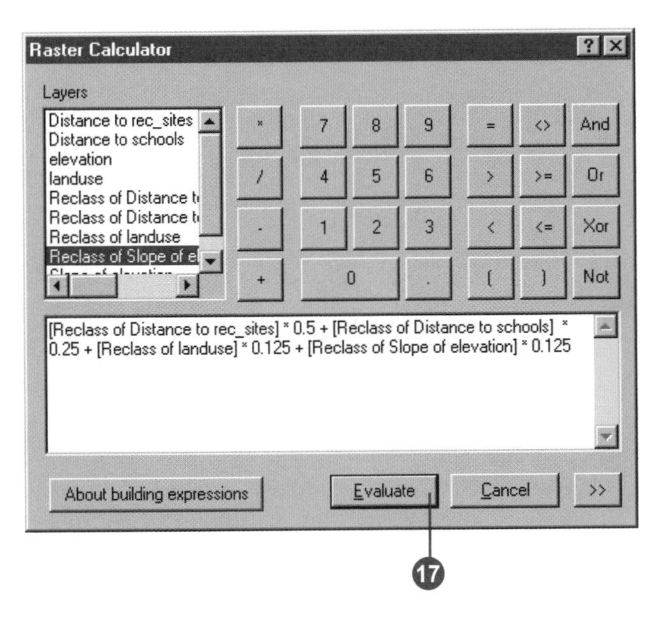

The output raster dataset shows you how suitable each location is for locating the new school, according to the criteria you set in the suitability model. Higher values indicate locations that are more suitable.

The suitable locations are those areas that are close to recreation sites, away from existing schools, on relatively flat land, and on certain types of landuse. The higher weightings set for Distance to schools and Distance to rec_sites have a powerful influence on deciding which areas are more suitable than others.

18. Right-click the newly created raster layer in the table of contents and click Properties.
19. Click the Symbology tab.
20. Click Classified from the Show list.
21. Click the Classes dropdown arrow and click 10.
22. Scroll to the last three classes, click one, then press and hold the Shift key and click the other two.
23. Right-click the highlighted classes, click Properties for selected Colors, and click a bright color.
24. Click the Display NoData as dropdown arrow and click the color black. This displays values of NoData—Water and Wetlands—in this color.
25. Click OK.

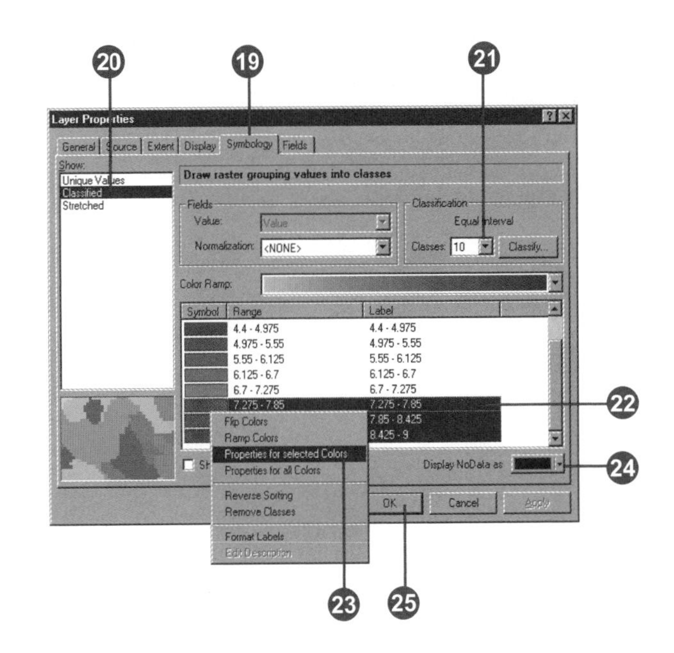

You decide that there are three main potential areas for locating the school. They are labeled in the diagram below.

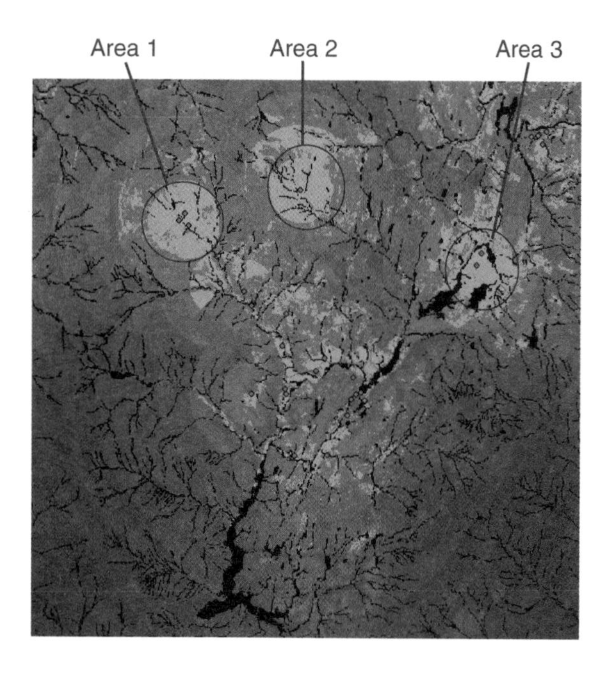

You should now assess these locations to see which might be the best location. This should be done in the field, as well as by examining the data you have on each potential area.

26. Right-click the output layer in the table of contents and click Make Permanent.
27. Navigate to the folder on your local drive where you set up your working directory (c:\spatial).
28. Type Suitability and click Save.

    The temporarily created dataset will now be permanently stored on disk.

    Note: A copy of this Suitability dataset can be found in the location ArcGIS\ArcTutor\Spatial\Results\Ex2\Suitability.

29. Click the output raster twice slowly. Rename the layer Suitability.

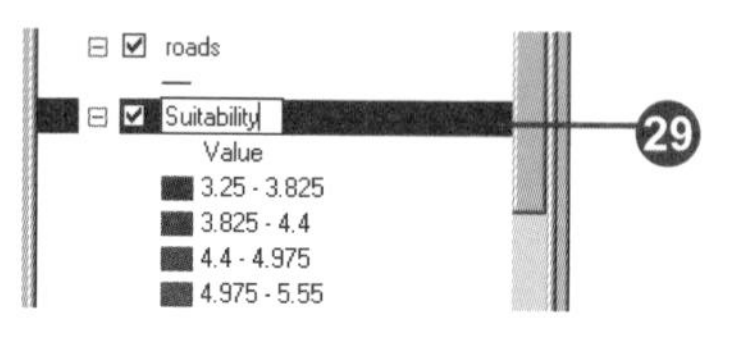

You decide that the best location is somewhere within Area 1, as there are three recreation sites in the neighboring area, the ski resort being one of them. Also, although you know that a considerable volume of traffic already uses the current access road to this potential site, you are involved in plans for constructing an alternative road to this area, which will help alleviate the volume of traffic on the current access road.

30. Click the Schools layer in the table of contents, press the Ctrl key, and click all other layers except Suitability (use the scroll bar to move down the table of contents).
31. Right-click one of the highlighted layers and click Remove.

You have now completed Exercise 2. You can continue on to Exercise 3, or you can stop and continue later. Whichever option you choose, save the map document at this point. Click the File menu and click Save As. Navigate to the location where you set up your local working directory (c:\spatial), specify a filename for the map document—Spatial_Tutorial—and click Save.

Exercise 3: Finding an alternative access road to the new school site

In this exercise you will find the best route for a new access road. The steps you might follow to produce such a path are outlined below, and the steps you will take in this exercise are diagrammed to the right.

Step 1: Create Source and Cost Datasets

Create the *source dataset* if necessary. The Source is the school site in this exercise.

Create the *cost dataset* by deciding which datasets are required, reclassifying them to a common scale, weighting, then combining them.

Step 2: Perform Cost Weighted Distance

Perform *cost weighted distance* using the Source and Cost datasets as inputs. The Distance dataset created from this function is a raster where the value of each cell is the accumulated cost of traveling from each cell back to the source.

To find the *shortest path*, you need a Direction dataset, which can be created as an additional dataset from the cost weighted function. This gives you a raster of the direction of the least costly path from each cell back to the source (in this exercise, the school site).

Step 3: Perform Shortest Path

Create the *destination* dataset if necessary. In this exercise, the Destination is a point at a road junction.

Perform shortest path using the Distance and Direction datasets created from the cost weighted function.

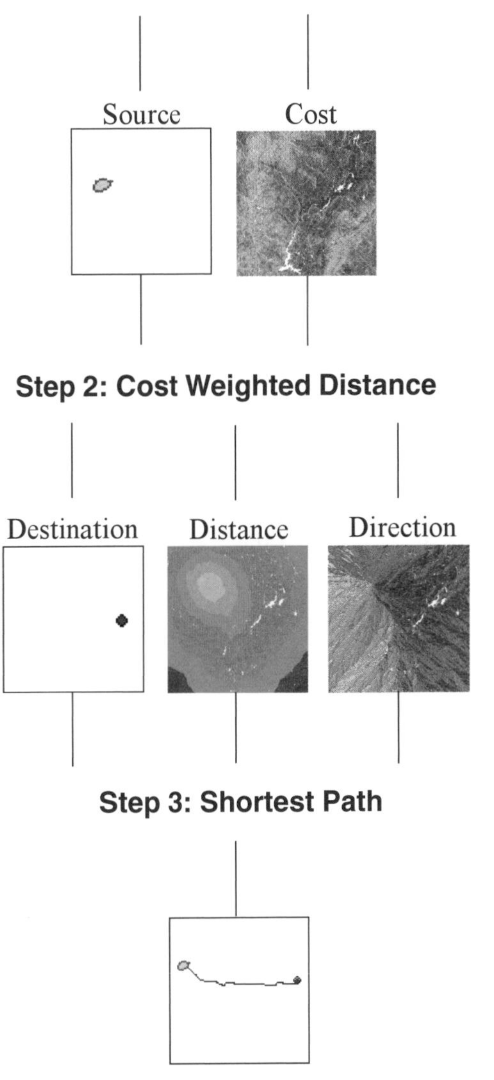

Step 1: Creating the source and cost datasets

To find the best route to the potential school site, you will first need to create the Source dataset (the school site) from the suitability map, and a Cost dataset, and use these as inputs into the cost weighted function.

Creating the source dataset

If you want to know how to create the Source dataset, follow the next 29 steps. Alternatively, click the Add Data button and navigate to the location where you installed the tutorial data (ArcGIS\ArcTutor\Spatial). Click Roads, then click Add. Then, click the Add Data button again and navigate to ArcGIS\ArcTutor\Spatial\Results\Ex3. Click School_site, then click Add and skip the next 29 steps.

You will first create an empty shapefile in ArcCatalog™, then digitize the location of the site using the editng tools in ArcMap.

1. Click the ArcCatalog button on the Standard toolbar.

2. Navigate in the *Catalog tree* to the folder on your local drive where you set up your working directory (c:\spatial).
3. Right-click the Spatial folder, point to New, and click Shapefile.

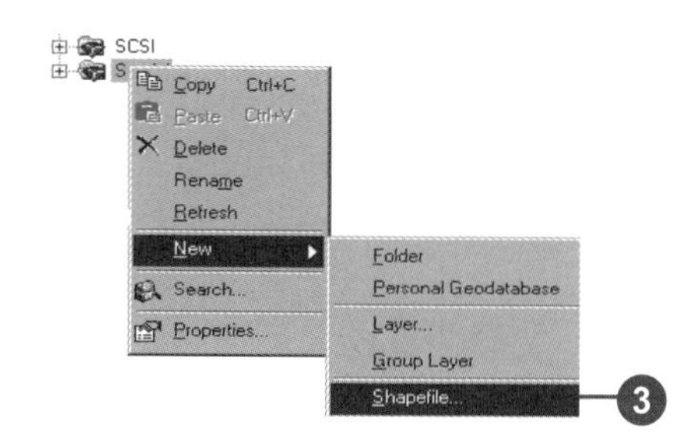

4. Type "School_site" for the name of the new shapefile.
5. Click the Feature Type dropdown arrow and click Polygon to choose the type of *feature* that will be created.
6. Click Edit to add *spatial reference* information to the shapefile.

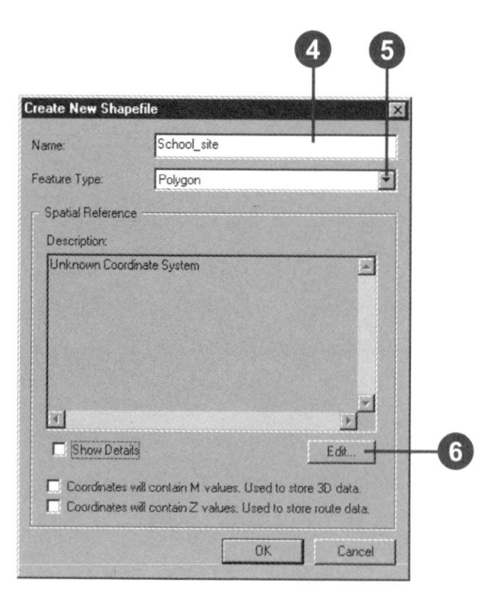

7. Click Select to use a predefined *coordinate system*.

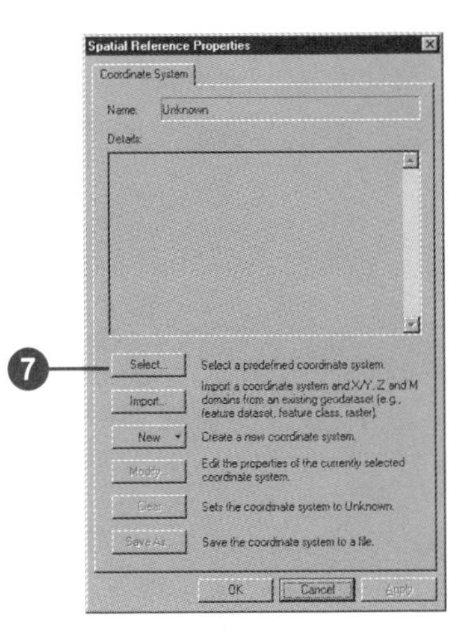

8. Click the Projected Coordinate Systems folder, click State Plane, then click NAD 1983 and scroll to NAD 1983 StatePlane Vermont FIPS 4400.prj.
9. Click Add.

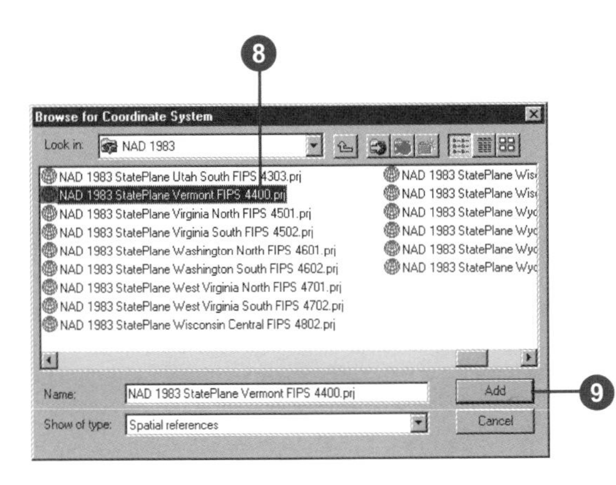

10. Click OK on the Spatial Reference Properties dialog box.
11. Click OK on the Create New Shapefile dialog box.

    A new shapefile called School_site will be created and added to the Catalog tree.
12. Click File, click Exit to close ArcCatalog, and return to ArcMap.
13. Click the Add Data button and navigate to the folder on your local drive where you installed the tutorial data (the default installation path is ArcGIS\ArcTutor\Spatial).
14. Click roads.shp.
15. Click Add.

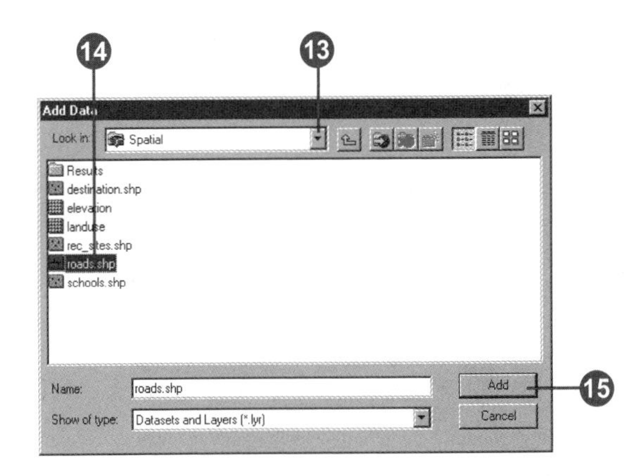

16. Click the Add Data button again and navigate to the folder on your local drive where you set up your working directory (c:\spatial).
17. Click School_site.
18. Click Add.

19. Click the Zoom In tool on the Tools toolbar and zoom in on the area that was deemed most suitable (area 1, circled in yellow below).

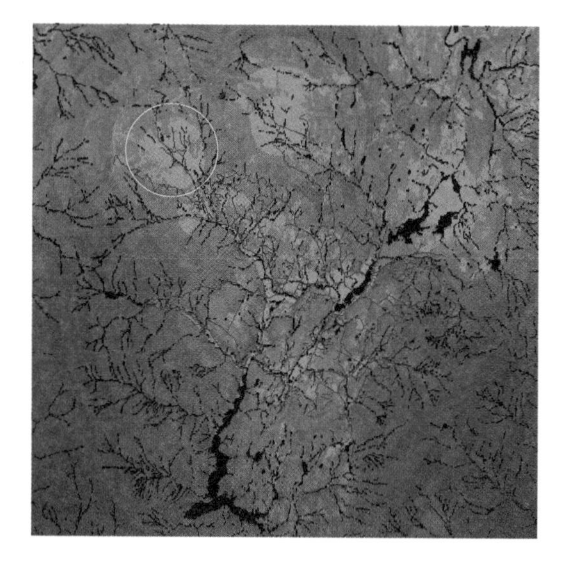

20. Click View, point to Toolbars, and click Editor.

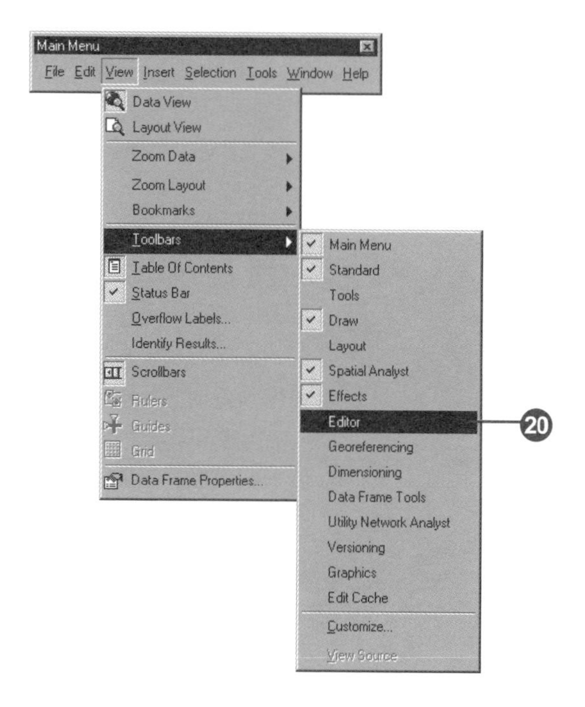

21. Click the Editor dropdown arrow and click Start Editing.

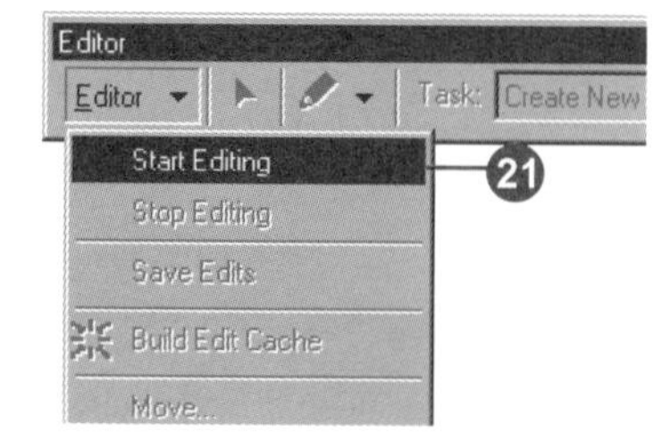

22. Click c:\spatial (or wherever you specified your working directory to be) for the folder from which to edit data.
23. Click OK.
24. Click the Task dropdown arrow and click Create New Feature.
25. Click the *Target* dropdown arrow and click School_site.
26. Click the Create New Feature dropdown arrow and click Create New Feature.

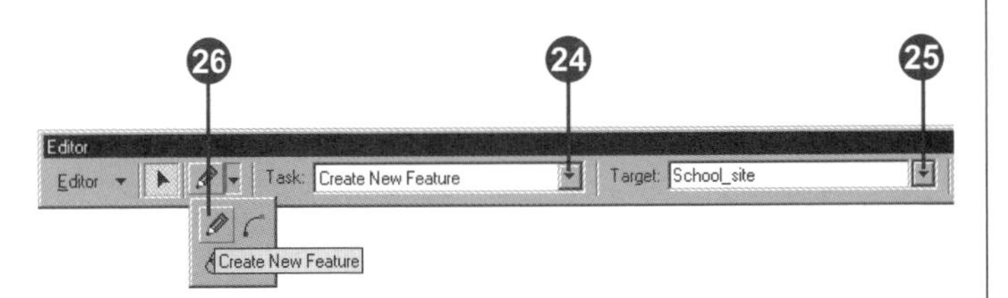

27. Draw a polygon on the screen in the location shown in the diagram. Click and hold to add a polygon vertex, drag the cursor, and add another polygon vertex. Continue until the polygon is complete. Double-click to close the polygon.

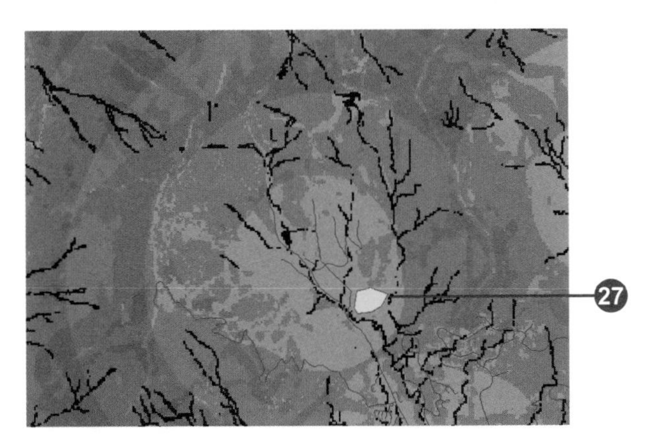

28. Click the Editor dropdown arrow and click Stop Editing.
29. Click Yes to save your edits.

Note: A copy of this school_site dataset can be found in the location ArcGIS\ArcTutor\Spatial\Results\Ex3\source.shp.

Creating the cost dataset

You will now create a dataset of the cost of traveling over the landscape, based on the fact that it is more costly to traverse steep slopes and construct a road on certain landuse types.

1. Right-click the Suitability layer and click Remove.
2. Click the Add Data button and navigate to the folder on your local drive where you set up your working directory (c:\spatial).
3. Click slope (the dataset created in exercise 2).
4. Click Add.
5. Click the Add Data button again and navigate to the folder on your local drive where you installed the tutorial data (the default installation path is ArcGIS\ArcTutor\Spatial).
6. Click landuse and click Add.

7. Right-click landuse and click Zoom To Layer.

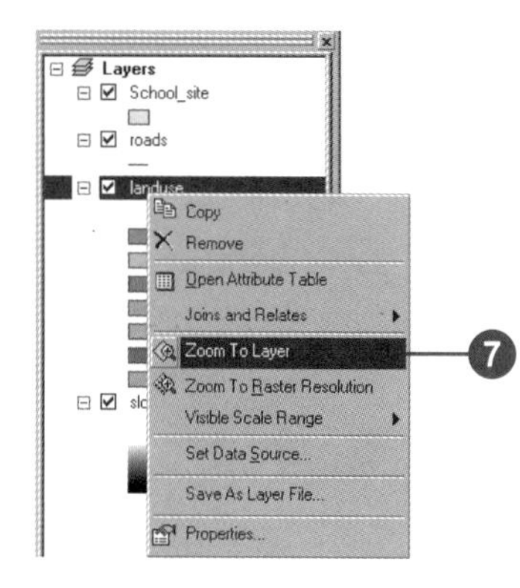

Reclassifying slope

1. Click the Spatial Analyst dropdown arrow and click Reclassify.

2. Click the Input raster dropdown arrow and click slope.
3. Click Classify.

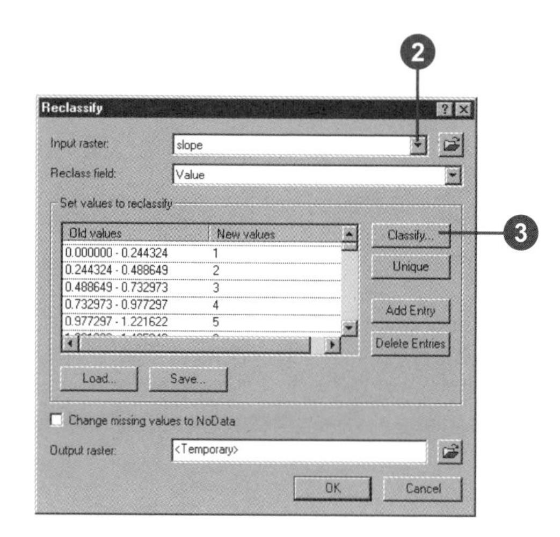

4. Click the Method dropdown arrow and click Equal Interval.
5. Click the Classes dropdown arrow and click 10.
6. Click OK.

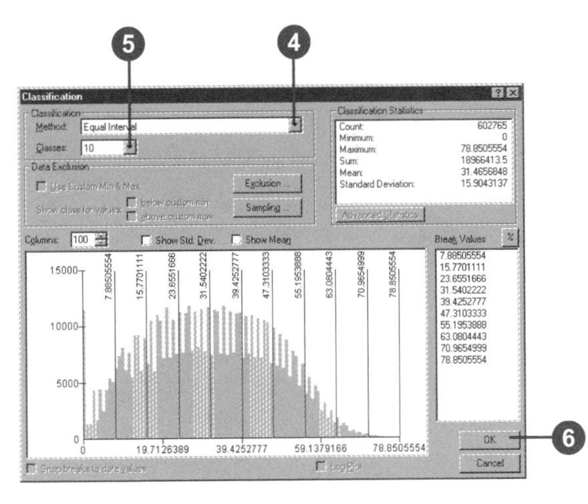

You want to avoid steep slopes when constructing the road, so steep slopes should be given higher values in the Cost dataset.

As the defaults give high values to steeper slopes, you do not need to change the default New Values.

7. Click OK on the Reclassify dialog box.

   The Reclass of slope layer will be added to the table of contents. It shows locations that are more costly than others for constructing a road—higher values indicate the more costly areas that should be avoided.

Reclassifying landuse

1. Click the Spatial Analyst dropdown arrow and click Reclassify.

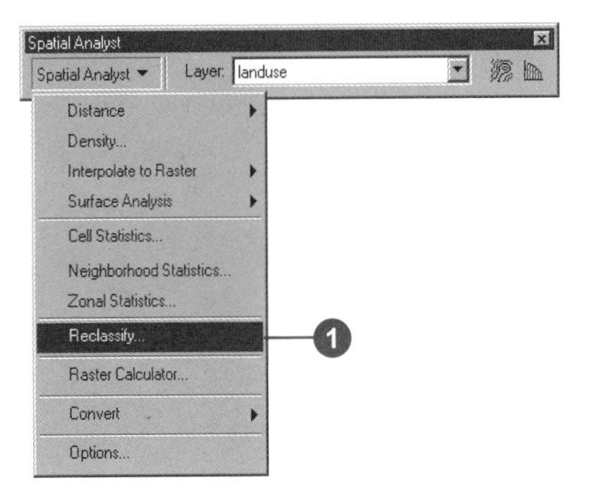

2. Click the Input raster dropdown arrow and click landuse.
3 Click the Reclass field dropdown arrow and click Landuse.
4. Click the first New value to edit the values and type in the following values:

Agriculture—4	Built up—9
Barren land—6	Forest—8
Brush/Transitional—5	Water—10

Higher values indicate higher road-construction costs.

5. Click Wetlands and click Delete Entries.
6. Click Change missing values to NoData.
7. Click OK.

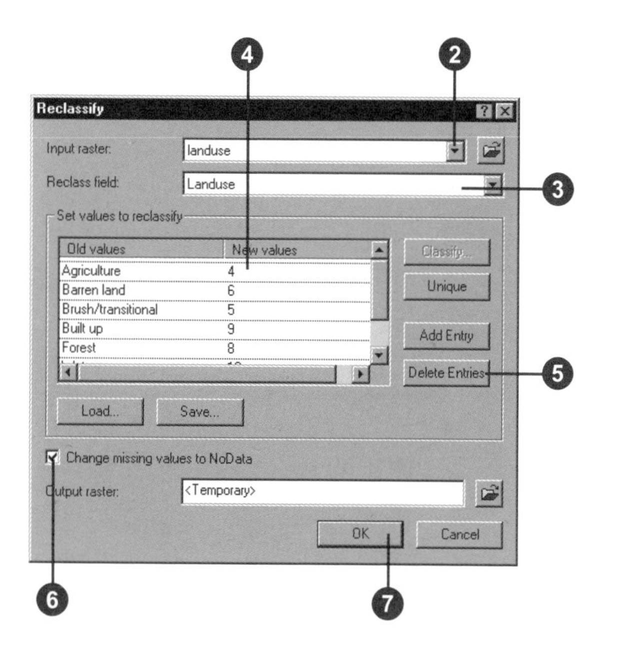

The Reclass of landuse layer will be added to the table of contents. It shows locations that are more costly than others for constructing a road, based on the type of landuse.

The NoData value—Wetlands—is currently displayed transparently so you can see the layers underneath. To make this value solid, change it to white.

8. Right-click Reclass of landuse and click Properties.
9. Click the Symbology tab.
10. Click the Display NoData as dropdown arrow and click Arctic White.
11. Click OK.

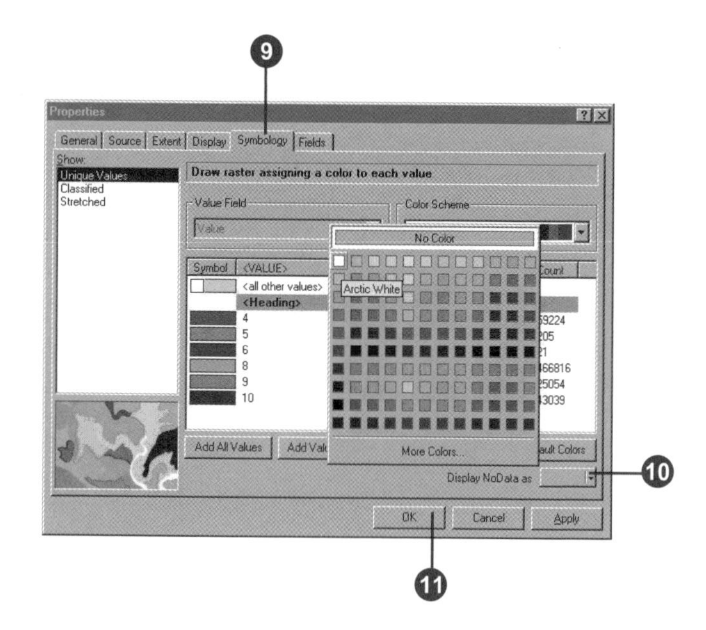

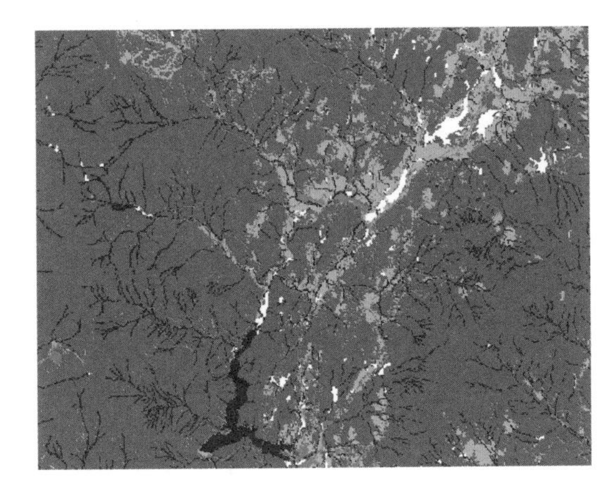

Combining datasets

You will now combine Reclass of slope and Reclass of landuse in order to produce a dataset of the cost of building a road at each location in the landscape, in terms of steepness of slope and landuse type. In this model, each dataset has equal weighting, so it is not necessary to apply any weight as we did when finding the suitable location for the school.

1. Click the Spatial Analyst dropdown arrow and click Raster Calculator.
2. Double-click Reclass of landuse to add it to the expression box.
3. Click the Add button.
4. Double-click Reclass of slope to add it to the expression box.
5. Click Evaluate.

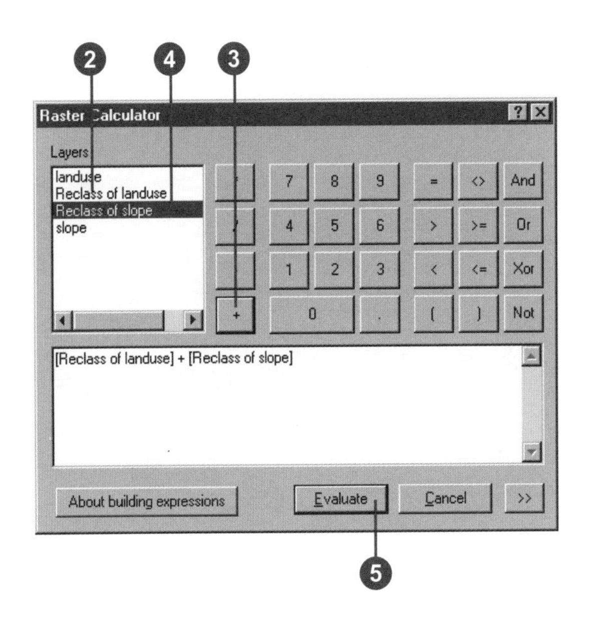

The result is added to your ArcMap session. Locations with low values identify the areas that will be the least costly to build a road through. They are displayed in dark blue in the graphic below.

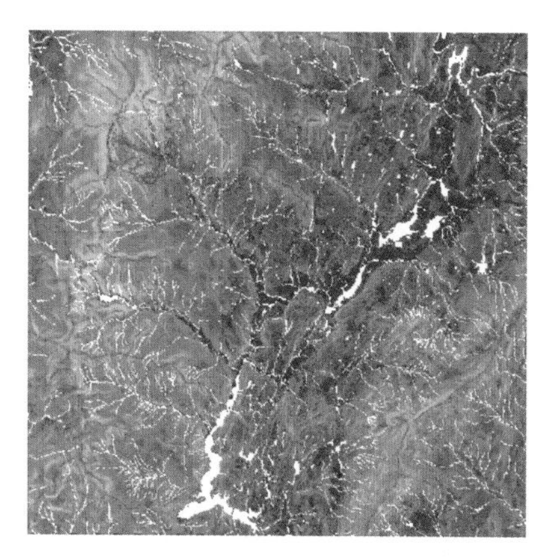

6. Click the output layer in the table of contents to highlight it, click again, and rename it Cost.

   You will now remove all layers except Cost, School_site, and Roads.

7. Click Reclass of landuse, press and hold down the Ctrl key, click Reclass of slope, slope, and landuse.
8. Right-click one of the layers and click Remove.

Step 2: Performing cost weighted distance

You will now perform cost weighted distance using the Cost dataset you just created and the School_site layer (the source). Using this function, you will create a Distance dataset where each cell contains a value representing the accumulated least cost of traveling from that cell to the school site and a Direction dataset that gives the direction of the least-cost path from each cell back to the source. This conceptual process is explained in more detail in Chapter 7, 'Performing spatial analysis'.

1. Click the Spatial Analyst dropdown arrow, point to Distance, and click Cost Weighted.

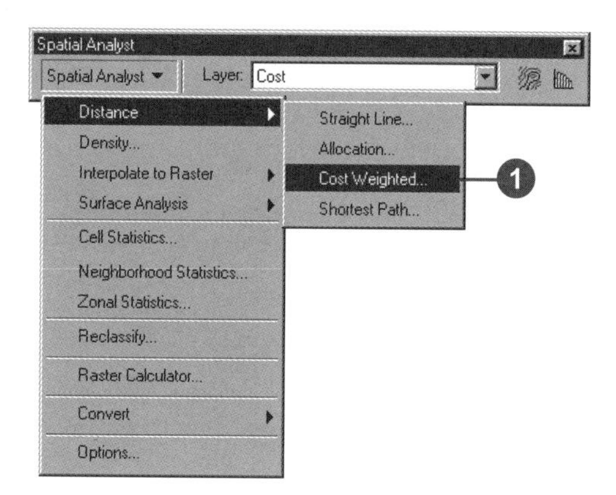

2. Click the Distance to dropdown arrow and click School_site.
3. Click the Cost raster dropdown arrow and click Cost.
4. Check Create direction.
5. Click OK.

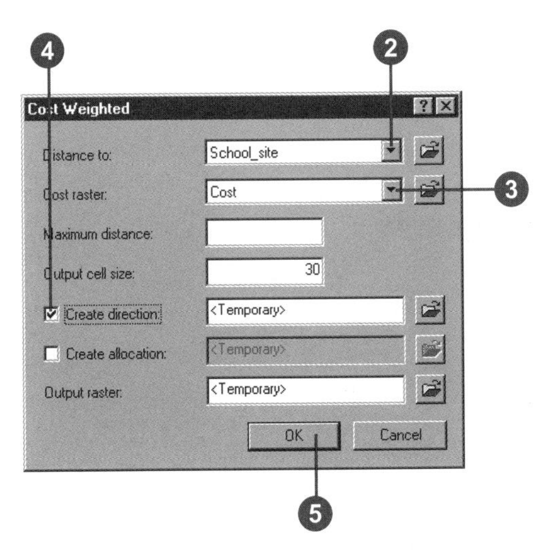

The Distance and Direction datasets are added to your ArcMap session as layers.

6. Click the output Distance layer in the table of contents, click again, and rename it Distance.
7. Click the output Direction layer in the table of contents, click again, and rename it Direction.

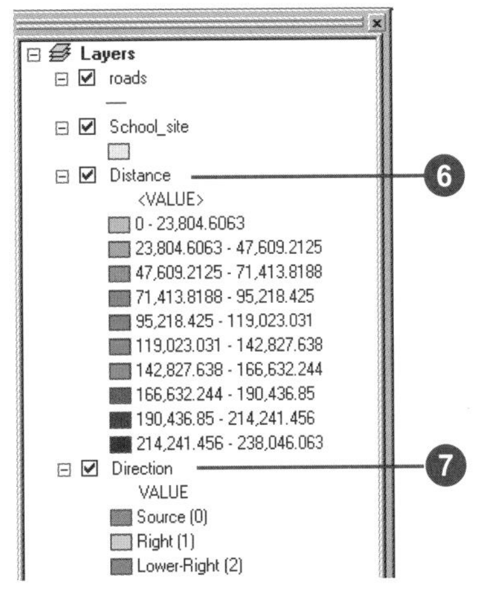

Step 3: Performing shortest path

You are now almost ready to find the shortest path from the school site. You have performed cost weighted distance, creating a Distance dataset and a Direction dataset, using the school site as the source. However, you will need to decide on, and then create, the destination point for the road. As you have already learned how to create a new shapefile, this destination point shapefile has been created for you.

1. Click the Add Data button.
2. Navigate to the folder on your local drive where you installed the tutorial data (the default installation path is ArcGIS\ArcTutor\Spatial).
3. Click Destination and click Add.

   Locating the destination point for the road in the position identified by the Destination shapefile will take much of the traffic away from the current road and provide a back route to the area for school buses and other vehicles.

4. Click the Spatial Analyst dropdown arrow, point to Distance, and click Shortest Path.

5. Click the Path to dropdown arrow and click Destination.
6. Click the Cost distance raster dropdown arrow and click Distance.
7. Click the Cost direction raster dropdown arrow and click Direction.

   Leave the defaults for the other options.
8. Click OK.

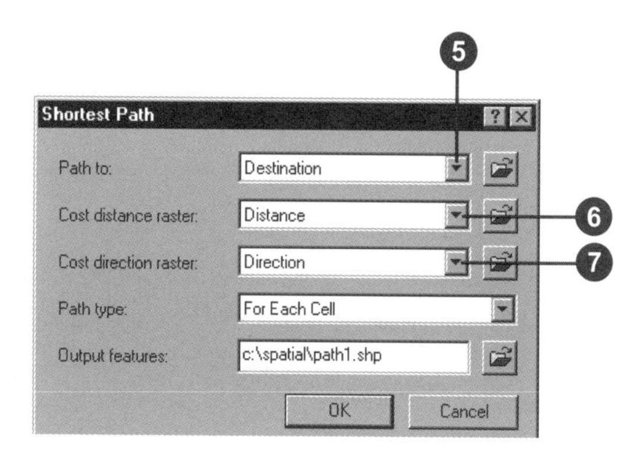

The shortest path is calculated, and the resulting layer is added to your ArcMap session. It represents the least-cost path—least cost meaning avoiding steep slopes and on landuse types considered to be least costly for constructing the road—from the school site to the road junction.

9. Click Distance in the table of contents, press the Ctrl key, and click Direction and Cost.
10. Right-click Cost and click Remove to remove all three layers.

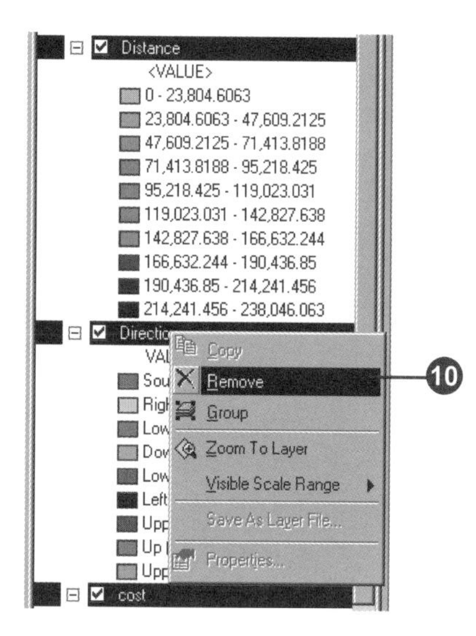

Displaying the results

To see exactly where this path should be constructed, you will now create a more detailed map.

Adding the datasets

1. Click the Add Data button on the Standard toolbar.
2. Navigate to the folder on your local drive where you set up your working directory (c:\spatial).
3. Click Hillshade and click Add.

   Note: A copy of this hillshade dataset can be found in the location ArcGIS\ArcTutor\Spatial\Results\Ex1\Hillshade.
4. Click the Add Data button on the Standard toolbar.
5. Navigate to the folder on your local drive where you installed the tutorial data (the default installation path is ArcGIS\ArcTutor\Spatial).
6. Click landuse and click Add.

Applying transparency

7. If the Effects toolbar is not already present, click View on the Main menu, point to Toolbars, and click Effects.
8. Click the Layer dropdown arrow on the Effects toolbar and click landuse.
9. Click the Adjust Transparency button and move the scroll bar up to a transparency of 30 percent.

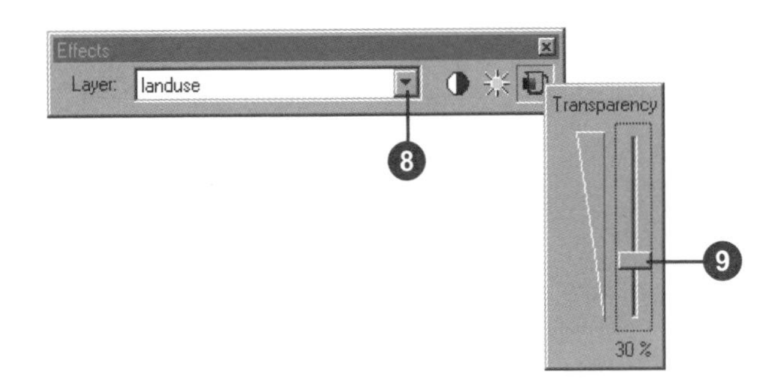

Changing the default field for landuse

You will now change the value field for the landuse layer so you can more easily distinguish each landuse type.

1. Right-click Landuse in the table of contents and click Properties.
2. Click the Symbology tab.
3. Click the Value Field dropdown arrow and click landuse.
4. Click OK.

   Change the color of the symbols in the table of contents to more appropriate colors for each type of landuse.

5. Right-click the symbols representing landuse types in the table of contents and pick an appropriate color for each one.

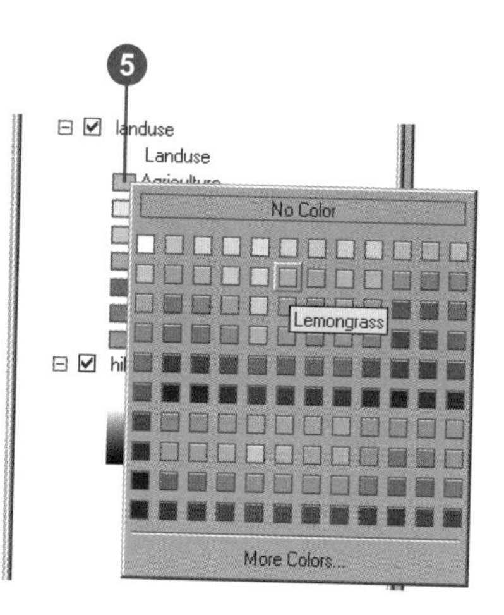

Zooming in on the area

1. Click the Zoom In tool on the Tools toolbar.

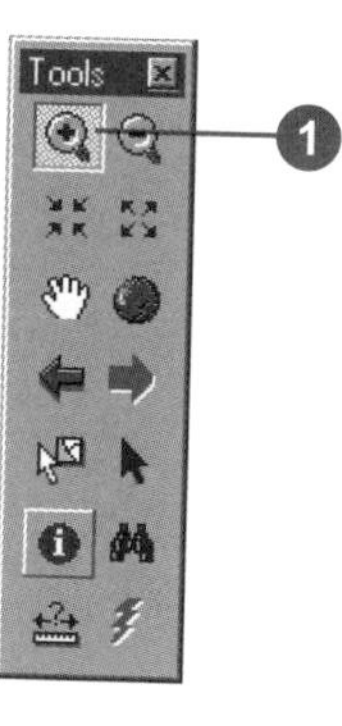

2. Click and drag a rectangle around the location of the new road to zoom in on this area (the area to zoom to is highlighted in red on the map below).

Labeling the roads

Label the road network to be able to identify which existing roads may be of use in constructing the new road.

1. Right-click Roads in the table of contents and click Properties.
2. Click the Labels tab.
3. Check Label Features.
4. Click the Label Field dropdown arrow and click STREET_NAM.
5. Click OK.

The road names are labeled on the map.

6. Click the File menu and click Save.

   If this is the first time you are saving the map document, navigate to the location where you set up your working directory (c:\spatial), specify a filename for the map document—Spatial_Tutorial—and click Save.

This brings you to the end of this tutorial. You have been introduced to some of the functions of Spatial Analyst, such as learning how to explore your data, producing a suitability map, and finding the least-cost path.

You have covered much ground, but there is a great deal more for you to explore. The rest of this book will provide you with a guide as you learn how to solve your own specific spatial problems.

Modeling spatial problems

3

IN THIS CHAPTER

- **Modeling spatial problems**
- **A conceptual model for solving spatial problems**
- **Using the conceptual model to create a suitability map**

Spatial Analyst can help you perform useful analysis, but it cannot solve problems by itself. To get the results you are hoping for, you have to ask the right questions and provide the right information. This chapter will introduce you to the concept of spatial modeling to help you recognize the conceptual steps involved in performing spatial analysis.

This chapter will explain:

- Modeling spatial problems.
- The conceptual modeling process:
    - Stating the problem
    - Breaking the problem down
    - Exploring input datasets
    - Performing analysis
    - Verifying the model's result
    - Implementing the result
- Following the conceptual modeling process to build a suitability model. The suitability model from Exercise 2 of the quick-start tutorial, 'Finding a site for a new school in Stowe, Vermont, USA', will be broken down conceptually to explain each of the modeling steps.

Modeling spatial problems

In general terms, a model is a representation of reality. Due to the inherent complexity of the world and the interactions in it, models are created as a simplified, manageable view of reality. Models help you understand, describe, or predict how things work in the real world. There are two main types of models: those that represent the objects in the landscape—representation models—and those that attempt to simulate processes in the landscape—process models.

Representation models

Representation models try to describe the objects in a landscape, such as buildings, streams, or forest. The way representation models are created in a GIS is through a set of data layers. For Spatial Analyst, these data layers will be either raster or feature data. Raster layers are represented by a rectangular mesh or grid, and each location in each layer is represented by a grid cell, which has a value. Cells from various layers stack on top of each other, describing many attributes of each location.

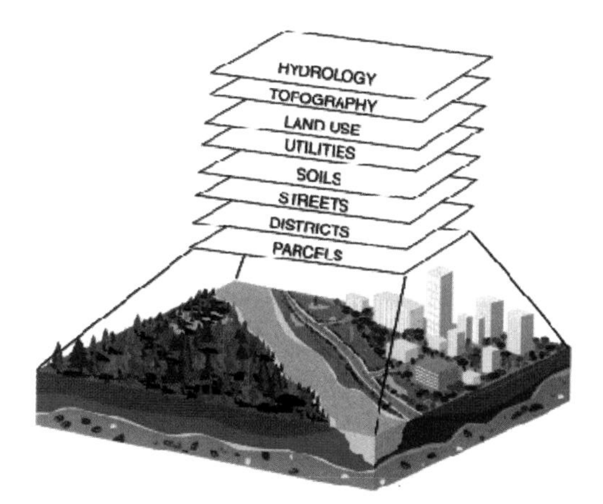

The representation model attempts to capture the spatial relationships within an object—the shape of a building—and between the other objects in the landscape—the distribution of buildings. Along with establishing the spatial relationships, the GIS representation model is also able to model the attributes of the objects—who owns each building. Representation models are sometimes referred to as data models and are considered descriptive models.

Process models

Process models attempt to describe the interaction of the objects that are modeled in the representation model. The relationships are modeled using spatial analysis tools. Since there are many different types of interactions between objects, ArcGIS and Spatial Analyst provide a large suite of tools to describe interactions. Process modeling is sometimes referred to as cartographic modeling. Process models can be used to describe processes, but they are often used to predict what will happen if some action occurs.

Each Spatial Analyst operation and function can be considered a process model. Some process models are simple, while others are more complex. Even more complexity can be added by including logic, combining multiple process models, and using the Spatial Analyst object model and Microsoft® Visual Basic®.

One of the most basic Spatial Analyst operations is adding two rasters together:

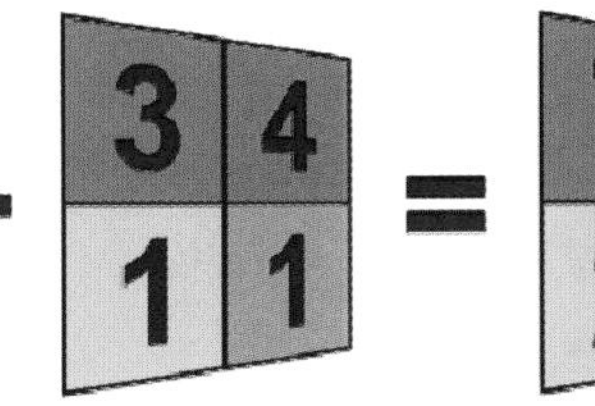

Complexity can be added through logic:

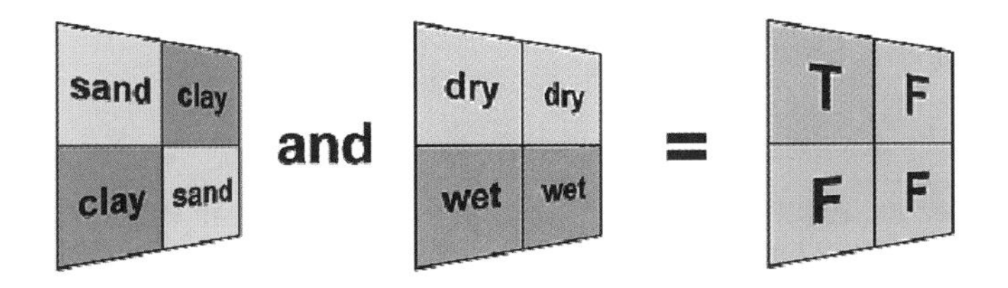

Additional complexity is added through specialized functions:

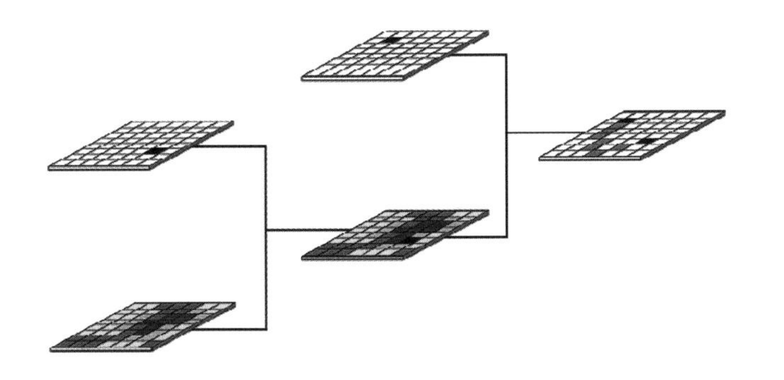

And even more complexity is added by combining several functions and logic:

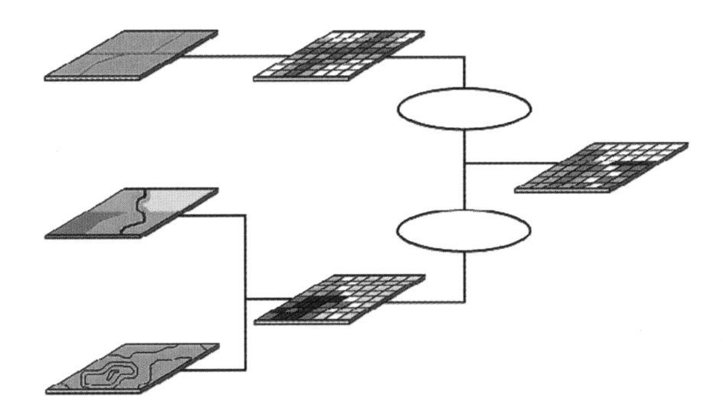

A process model should be as simple as possible to capture the necessary reality to solve your problem. You may just need a single operation or function, but sometimes hundreds of operations and functions may be necessary.

Types of process models

There are many types of process models to solve a wide variety of problems. Some include:

- **Suitability modeling:** Most spatial models involve finding optimum locations, such as finding the best location to build a new school, landfill, or resettlement site.
- **Distance modeling:** What is the flight distance from Los Angeles to San Francisco?
- **Hydrologic modeling:** Where will the water flow to?
- **Surface modeling:** What is the pollution level for various locations in a county?

A set of conceptual steps can be used to help you build a model. The remainder of this chapter explains these steps.

A conceptual model for solving spatial problems

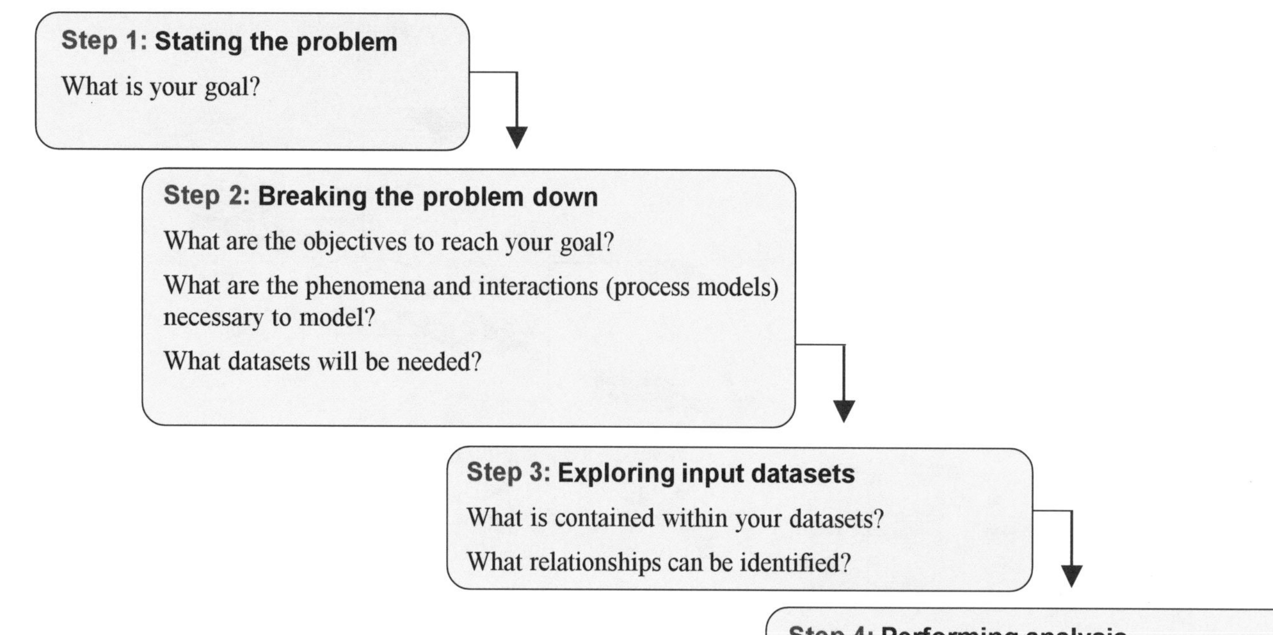

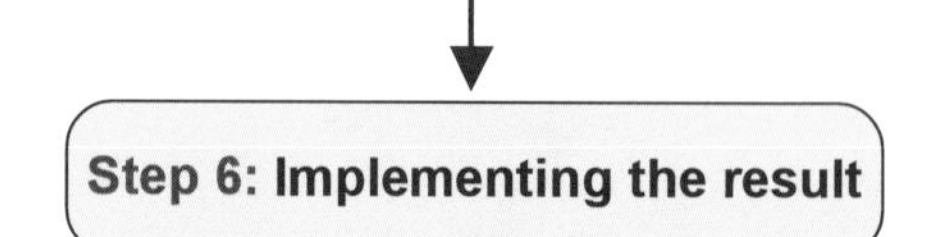

Step 1: Stating the problem

To solve your spatial problem, you need to start off by clearly stating the problem you are trying to solve. What is your goal? Following the steps below will help you realize your goal.

Step 2: Breaking the problem down

Once the goal of the problem is understood, you must then break the problem down into a series of objectives, identify the elements and their interactions that are needed to meet your objectives, and create the necessary input datasets to develop the representation models.

By breaking the problem down into a series of objectives, you will discover the necessary steps to reach your goal, which will help you to solve the problem. If your goal was to find the best sites for spotting moose, your objectives might be to find out where moose were recently spotted, what vegetation types they feed on most, and so on. By arranging the objectives in order, you will begin to understand the big picture of what you are ultimately trying to solve.

Once you have established your objectives, you need to identify the elements, and the interactions between these elements, that will meet your objectives. The elements will be modeled through representation models and their interactions through process models. Moose and vegetation types will be only a few of the elements necessary for identifying where moose are most likely to be. The location of humans and the existing road network will also influence the moose. The interactions between the elements are that moose prefer certain vegetation types and they avoid humans, who can gain access to the landscape through roads. A series of process models might be needed to ultimately find the locations with the greatest chance of spotting a moose.

During this step, you should also identify the necessary input datasets. Input datasets might contain sightings of moose in the past week, vegetation type, and the location of human dwellings and roads. Once you have identified them, they need to be represented as a set of data layers—a representation model. To do this, you need to understand how raster data is represented in Spatial Analyst. Chapter 4, 'Understanding raster data', explains the concepts involved when representing data.

The overall model—made up of a series of objectives, process models, and input datasets—provides you with a model of reality, which will help you in your decision making process.

Step 3: Exploring input datasets

It is useful to understand the spatial and attribute relationships of the individual objects in the landscape and the relationships between them—the representation model. To understand these relationships, you need to explore your data. A wide variety of tools are available in ArcGIS and Spatial Analyst to explore your data, and these tools are covered throughout the various books accompanying ArcGIS.

Step 4: Performing analysis

At this stage, you need to identify the tools to use to build the overall model. Spatial Analyst provides a wide variety of tools to serve this purpose. In our moose spotting example, you may need to identify the tools necessary to select and weight certain vegetation types and buffer houses and roads and weight them appropriately. Chapter 5, 'Understanding cell-based modeling', presents the principles for performing cell-based modeling and the issues that must be considered. Chapter 6, 'Setting up your analysis environment', and Chapter 7, 'Performing spatial analysis', show how these principles are realized in Spatial Analyst.

Step 5: Verifying the model's result

Check the result from the model in the field. Do certain parameters need changing to give you a better result?

If you created several models, determine which model you should use. You need to identify which model is best. Does one particular model clearly meet your initial goal better than the rest?

Step 6: Implementing the result

Once you have solved your spatial problem, verifying that the result from a particular model meets your initial expectations outlined in step 1, implement your result. When you visit the locations with the greatest chance of spotting moose, do you in fact see any?

Using the conceptual model to create a suitability map

Step 1: Stating the problem

To solve a spatial problem, you should first state the problem you are trying to solve. What is your goal? Start with a concept of the intended output of the study; visualize the type of map you want to produce.

To understand the step process, you will work through a sample problem for the remainder of this chapter. Your problem is to find the best location for siting a new school. The result you seek is a map showing potential sites—ranked best to worst—that could be suitable for building a new school. This is called a ranked suitability map because it shows a relative range of values indicating how suitable each location is on the map, taking into account the criteria you put into the model.

Best site for a new school

To help you model your spatial problem, draw a diagram of the steps involved:

Start with the statement of the problem. As you work through the problem, you will expand the diagram to show objectives, process models, and necessary input datasets to use to reach your goal.

Step 2: Breaking down the problem

Once the problem is stated, break it down into smaller pieces until you know what steps are required to solve it. These steps are objectives that you will solve.

When defining objectives, consider how you will measure them. How will you measure what is the best area for the new school? In siting the school, it is preferable to find a location near recreational facilities, as many of the families who have relocated to the town have young children interested in pursuing recreational activities. It is also important to be away from existing schools to spread their locations over the town. The school must also be built on suitable land that is relatively flat.

The graphic below outlines the objectives:

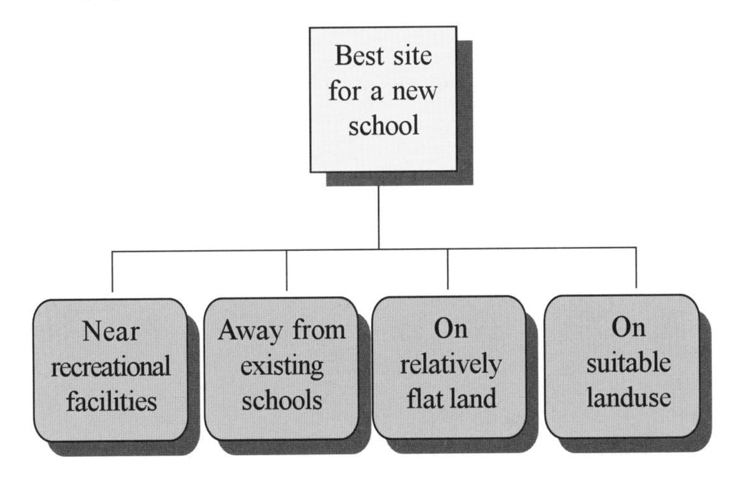

You want to know the following: "Where are locations with relatively flat land? Is the landuse in these locations of a suitable type? Are these locations close enough to recreation sites? Are they far enough away from existing schools?"

Are these locations close enough to recreation sites?

You know that it is preferable to locate the school close to recreational facilities, so you need to create a map displaying the distance to recreation sites to locate the school in areas that are close to them. The process model here involves calculating the distance from recreation sites.

Input dataset needed: location of recreational facilities

Are they far enough away from existing schools?

You want to site the school away from existing schools to avoid encroaching on their catchment areas. So you need to create a map displaying the distance to schools. Here, the process model involves calculating the distance from existing schools.

Input dataset needed: location of existing schools

Where are locations with relatively flat land?

To find areas of relatively flat land, you need to create a map displaying the slope of the land. The process model here involves calculating the slope of the land.

Input dataset needed: elevation

Is the landuse in these locations of a suitable type?

You need to decide what makes a suitable landuse type on which to build. This is a subjective process, according to your problem. Here, agricultural land is considered the least costly to build on and therefore the most preferable. Barren land is next, then brush–transitional, forest, and existing built-up areas. There is no process model involved here, just an identification of the input landuse dataset and which landuses are most preferable to build on.

Input dataset needed: landuse

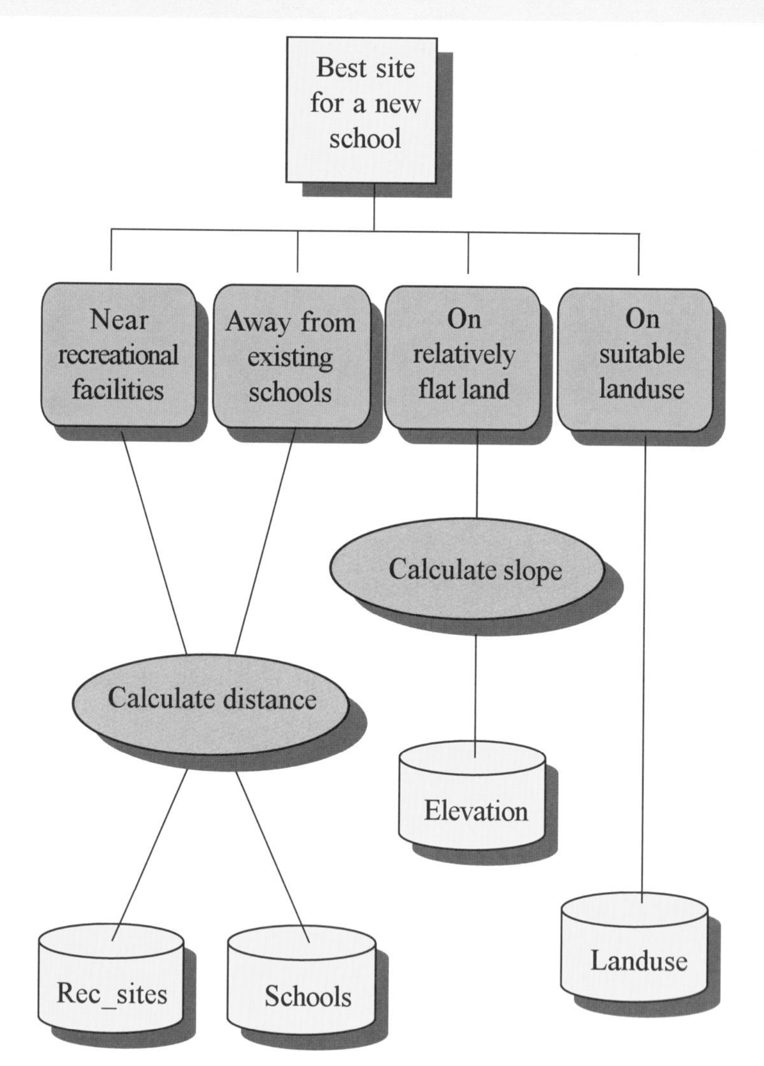

Step 3: Exploring input datasets

Once you have broken down your problem into a series of objectives and process models and decided what datasets you will need, you should explore your input datasets to understand their content. This involves understanding which attributes within and between datasets are important for solving the problem and looking for trends in the data.

By exploring your data, you can often gain insights about the areas you wish to locate the school in, the weightings for input attributes, and alterations to your modeling process. You can see the locations of existing schools and recreation sites, and you can tell from the elevation dataset where the higher elevations are. The landuse dataset tells you what types of landuse are in the area and where they are located in relation to the other datasets.

See Exercise 1 of the Quick-start tutorial for how to use some of the tools of ArcMap and Spatial Analyst to explore your data.

Identify features to get information from all layers.

Identify Results

Layers: <All layers>

- rec_sites
  - Golden Eagle Resort
- roads
  - PALISADES LN
- elevation
  - 769
- landuse
  - 4
- Hillshade of elevation
  - 221

Location: (484848.701952 218993.362001)

Field	Value
FID_1	12
Shape	Point
FID	307
AREA	0
PERIMETER	0
MAP	90013
SITE_NAME	Golden Eagle Resort
FIPS_CODE	15040
ALT_NAME	
S_ADDRESS	Mountain Road
S_TOWN	Stowe, VT 05672
S_PHONE	802-253-4811

Examine the attribute table for each layer.

Attributes of landuse

ObjectID*	Value	Count	Landuse
1	1	294	Brush/transitional
2	2	62187	Water
3	3	28	Barren land
4	4	36034	Built up
5	5	85054	Agriculture
6	6	671722	Forest
7	7	12241	Wetlands

Record: 1 Show: All Selected Records

Create and examine histograms from each layer.

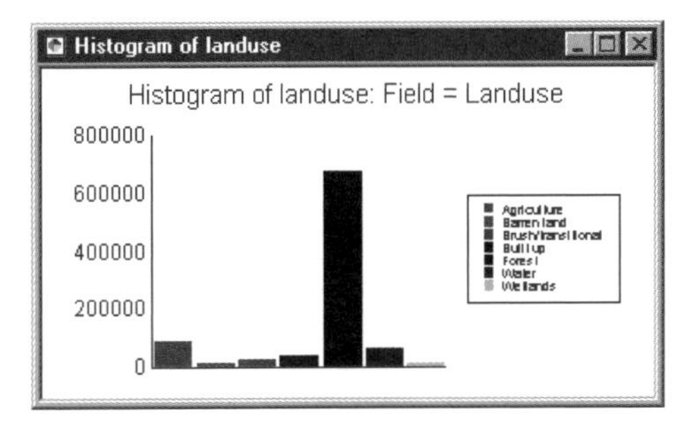

Calculate hillshade to examine the relief.

Step 4: Performing analysis

You have decided on your objectives, the elements and their interactions, the process models, and what input datasets you will need. You are now in the position to perform analysis.

The ESRI Guide to GIS Analysis describes in detail the many tasks that can be solved with ArcGIS.

When finding the best location for the new school, there are two ways to go about performing analysis. You can create a suitability map to find out the suitability of every location on the map, or you can simply query your created datasets to obtain a Boolean result of true or false.

Creating a suitability map

Creating a suitability map enables you to obtain a suitability value for every location on the map.

Once you have created the necessary layers, how do these created layers get combined to create a single ranked map of potential areas to site the school? You need a way to compare the values of classes between layers. One way to do this is to assign numeric values to classes within each map layer.

Each map layer is ranked by how suitable it is as a location for a new school. You may, for instance, assign a value to each class in each layer on a scale of 1–10, with 10 being the best. This is often referred to as a suitability scale.

NoData can be used to mask off areas that should not be considered. Having all measures on the same numeric scale gives them equal importance in determining the most suitable locations. The model is initially constructed in this way, then while testing alternative scenarios, weight factors can be applied to layers to further explore the data and its relationships.

Creating suitability scales

As is the case with this example, many scales are synthetic. These are often a ranked measure of suitability, or preference, from best to worst. It is based on something you can measure such as distance to schools, but in the end it is a subjective measure of how suitable a certain distance is from a school for locating another school.

There are natural scales that are commonly associated with some objectives. Cost is a good example but needs to be defined in sufficient detail. In a study of building suitability, an objective of low real estate cost would be measured on a scale of dollars. Be sure to adequately define the scale. For something as well understood as dollars, there are other variables such as whether it's U.S. dollars, Australian dollars, or an exchange rate between monies.

Many scales are not linear relationships, although they are often presented that way to save time and money or because all options were not considered. For example, if assigning a scale to travel distance, traveling 1, 5, or 10 kilometers would not be ranked as a suitability of 10, five, and one if you were walking. Some people may think walking 5 kilometers is only two times as bad as 1 kilometer, while others may think it's 10 times as bad.

When you construct a suitability scale, work with experts to find the best and worst of a scenario and as many intermediate points as possible. Experts should be knowledgeable about the objective being studied. For example, it is more meaningful to ask commuters to rank their opinions on drive-time desirability than to ask a city official when he thinks traffic is worst.

Ranking the areas close to recreation sites

To site the school close to recreational facilities, you need to know the distance to them. The Spatial Analyst Straight Line Distance function will create such a map, calculating the straight line distance from any location to the nearest recreation site. The result is a raster dataset in which every cell represents the distance to the nearest recreation site. To rank this map, simply use the Reclassify function. As it is preferable to locate close to recreation sites, give a value of 1 to distances far from recreation sites and a value of 10 to distances close to recreation sites, then rank the distances linearly in between as the following chart shows.

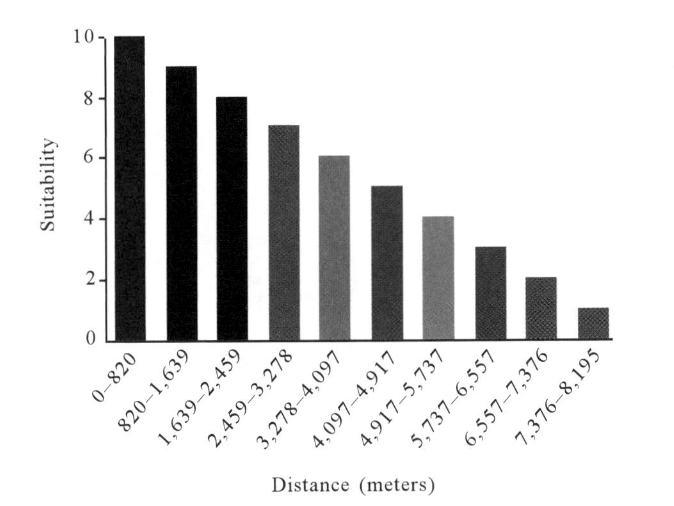

Ranking the areas away from existing schools

To avoid the catchment areas of the other schools, you need to know the distance to them. The Spatial Analyst Straight Line Distance function will create such a map, calculating the straight line distance from any location to the nearest school. The result is a raster dataset in which every cell represents the distance to the nearest school. To rank this map, simply use the Reclassify function. As it is preferable to locate away from existing schools, give a value of 1 to distances close to existing schools and a value of 10 to distances far from existing schools, then rank the distances linearly in between as the following chart shows.

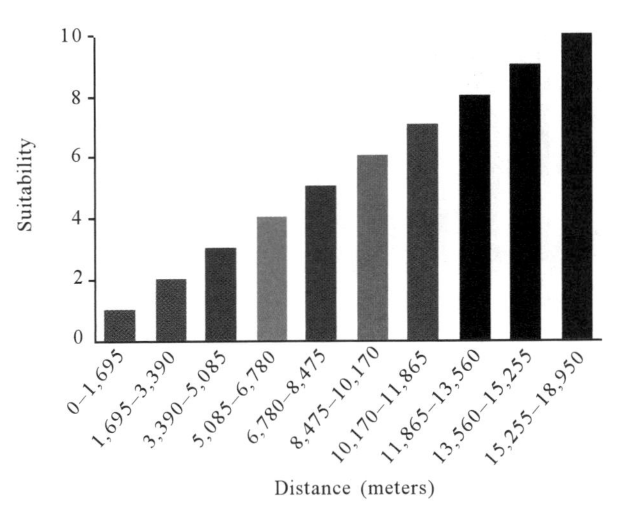

Ranking the areas on relatively flat land

To avoid steep slopes and find areas that are relatively flat to build on, you need to know the slope of the land. The Spatial Analyst Slope function will create such a map, identifying for each cell the maximum rate of change in value from each cell to its neighbors. To rank this map, simply use the Reclassify function. As it is preferable to locate on relatively flat slopes, give a value of 1 to locations with steep slopes, 10 to locations with the least steep slopes, then rank the values linearly in between as the following chart shows.

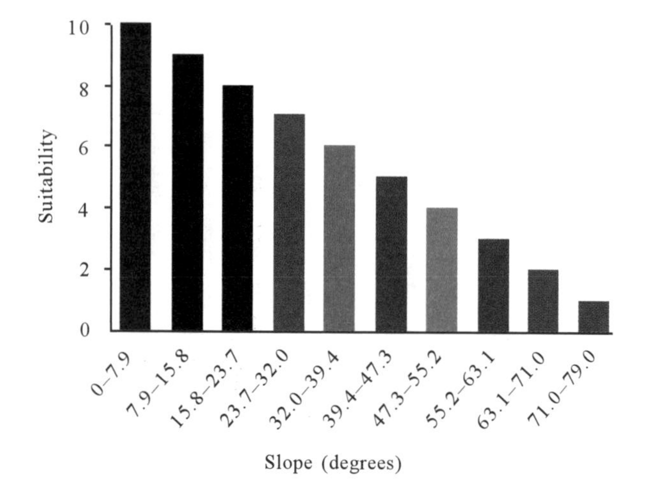

Ranking the areas on suitable landuse types

To rank the map representing landuse types, use the Reclassify function. As it is preferable to build on certain landuse types due to the costs involved, you need to decide how to rank the values.

Ranking distance or slope values is relatively straightforward. You simply have to decide whether short or long distances are preferable and whether steep slopes or less steep slopes are preferable, then rank the rest of the values linearly, or specify a maximum distance or slope to consider. Here you have to decide which landuse types are preferable. This is subjective depending on your study. The easiest way to decide what type of land is preferable for building on and what is not is to decide on the most preferable and then the least preferable. Then, out of the landuse types left, again decide on the most and least preferable. Do this until you have put the landuse type in order of preference. Landuses of Water and Wetlands have been excluded from the analysis since you cannot build on water, and there are restrictions against building on wetlands. The chart below shows how the landuse types have been ranked.

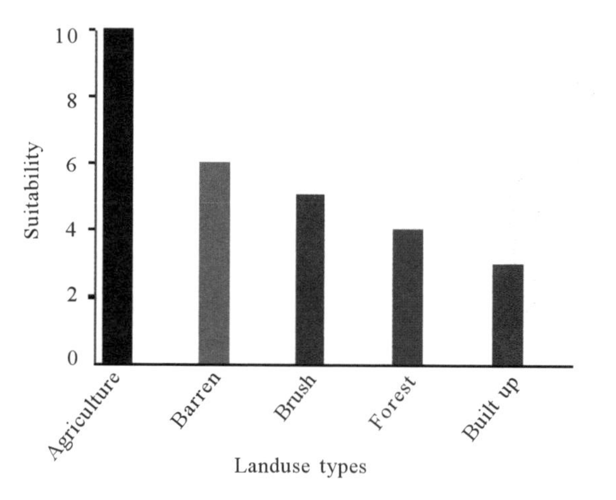

Combining the suitability maps

The last step in the suitability model is to combine the suitability maps of distance to recreation sites, distance to schools, slope, and landuse.

If all objectives had equal weight, the suitability maps could simply be combined at this point using the Raster Calculator. However, you know from breaking down the problem that the most preferable objective to satisfy is to locate the school close to recreational facilities, and the next is to locate away from existing schools.

To account for the fact that some objectives have more importance in the suitability model, you can weight the datasets, giving those datasets that should have more importance in the model a higher percentage influence—weight—than the others.

The following percentage influences will be assigned to the suitability maps. The values in brackets are the percentage divided by 100 to normalize the values. This normalized value will be assigned to each suitability map:

Distance to recreation sites:	50%	(0.5)
Distance to schools:	25%	(0.25)
Slope:	12.5%	(0.125)
Landuse types:	12.5%	(0.125)

So, the Distance to recreation sites suitability map has an influence of 50% (0.5) on the final result, and Distance to schools has an influence of 25% (0.25). Slope and Landuse types both have a 12.5% (0.125) influence. Like assigning scales of suitability, assigning weights is a subjective process, depending on what objectives are most important to your study.

The following graphs show the effect of applying the above weights on each suitability map.

Weights assigned to each suitability map

Notice how the values of suitability have changed by applying weights. For example, the suitability value for Agriculture was 10 in the original suitability map. By applying a weight of 0.125—or a percentage influence of 12.5%—the suitability value for Agriculture is now only 1.25. When these four weighted suitability maps are combined, the suitable locations for the school will have been influenced by the assigned weights. Areas close to recreation sites will have the most influence on the final suitability map.

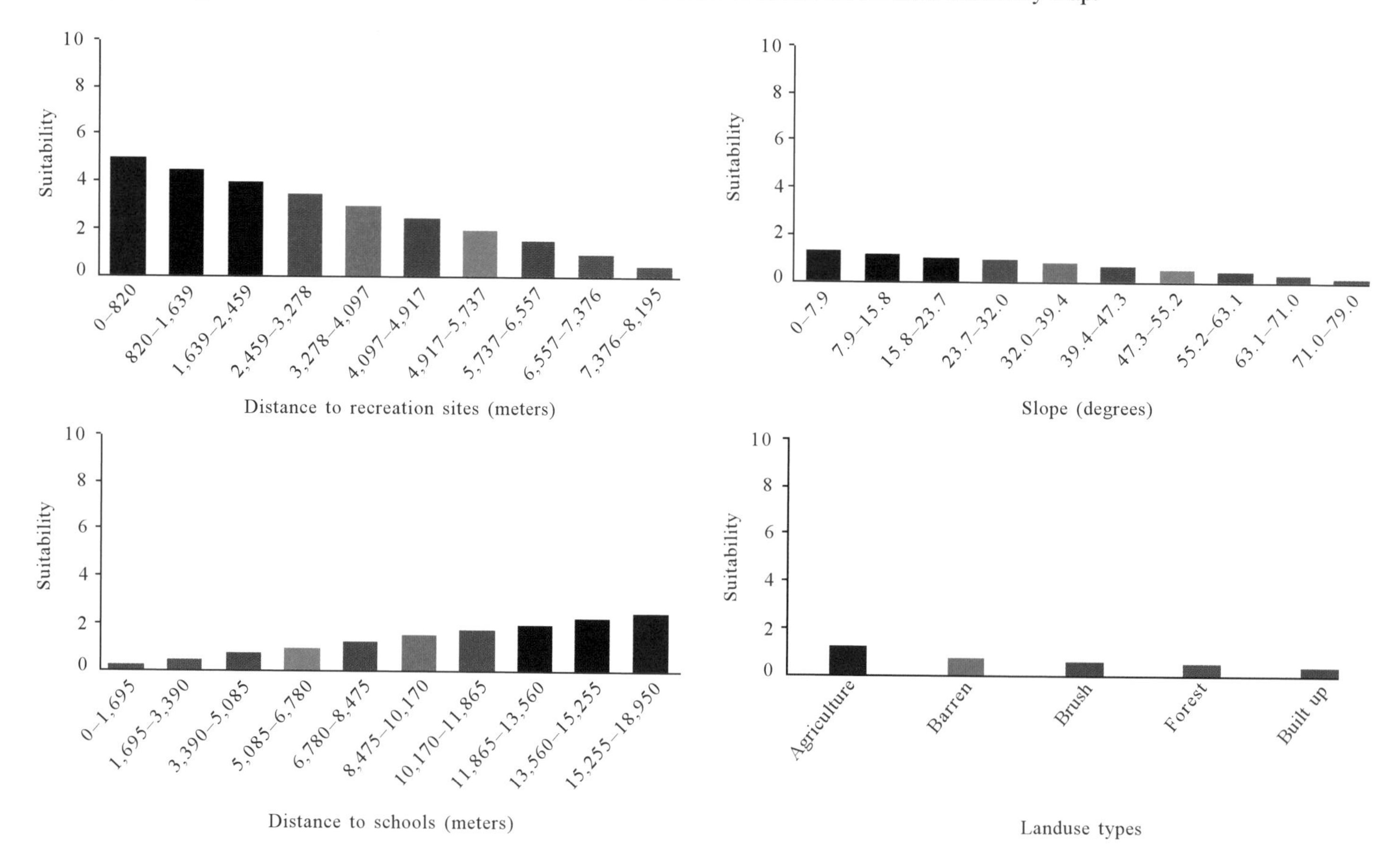

The final suitability map is produced by combining all the maps together. Weights can be assigned in the Raster Calculator at the same time as combining the suitability maps:

For example:

Distance to rec_sites * 0.5 + Distance to schools * 0.25 + Slope * 0.125 + Landuse * 0.125

The result will be a suitability map displaying the best locations for the new school. Higher values indicate more suitable locations.

See Exercise 2 of the quick-start tutorial for how to use Spatial Analyst to find the best location for the new school.

See Exercise 3 of the quick-start tutorial for how to use Spatial Analyst to find an alternative access road to the new school site.

Querying your data

The alternative way to find suitable locations for the new school—rather than creating a suitability map—is to query your data. Once you have created all the datasets you need—slope, distance to recreation sites, and distance to schools—you can simply query the data to find the suitable locations. Such a query would be to find all locations on agricultural land with slopes less than 20 degrees where the distance to recreation sites is less than 1,000 meters and the distance to schools is greater than 4,000 meters.

The above query in the Raster Calculator:

[landuse] == 5 & [Slope] < 20 & [Distance to rec_sites] < 1000 & [Distance to schools] > 4000

The result would give a Boolean true or false map of locations that meet or do not meet the criteria.

Step 5: Verifying the model's result

Once you have your result from any spatial analysis, you should verify that it is correct. This should be done, if possible, by visiting the potential sites in the field. Often the result you achieve has not accounted for something important, for instance, there may be a cow barn upwind of the site, producing foul odors, or by examining the town hall records you may discover a restriction on building on the desired land of which you were not aware. If either is the case, then you will need to add this information to the analysis.

Step 6: Implementing the result

The final step in the spatial model is to implement the result, building the new school in the chosen location.

Understanding rasters and analysis

Section 2

Understanding raster data

4

IN THIS CHAPTER

- **Understanding a raster dataset**
- **Coordinate space and the raster dataset**
- **Discrete and continuous data**
- **The resolution of a raster dataset**
- **Raster encoding**
- **Representing features in a raster dataset**
- **Assigning attributes to a raster dataset**
- **Using feature data directly in Spatial Analyst**
- **Deriving raster datasets from existing maps**

When using Spatial Analyst for some or all of your processing, you will have to use or create raster datasets. In this chapter you will be exposed to how a raster dataset is represented in Spatial Analyst and the issues you need to be aware of when using and creating rasters. This chapter will focus on the concerns of the raster representation, while Chapter 5, 'Understanding cell-based modeling', a companion chapter to this one, will address the issues that must be considered when performing analysis.

From this chapter you will learn:

- About the structure of raster datasets
- The importance of coordinate space and raster datasets
- The difference between discrete and continuous types of raster datasets
- About the resolution or cell size when creating a raster dataset
- How raster datasets are encoded and how points, lines, and polygons are represented as cells
- Other issues you need to be aware of, such as when adding other attributes to raster datasets and creating raster datasets from existing maps

Understanding a raster dataset

Raster data is generally divided into two categories: thematic data and image data. The values in thematic raster data represent some measured quantity or classification of a particular phenomena such as elevation, pollution concentration, or population. For example, in a landcover map the value 5 may represent forest, and the value 7 may represent water. The values of cells in an image represent reflected or emitted light or energy such as that of a satellite image or a scanned photograph. The analysis tools of Spatial Analyst are primarily intended for use on thematic raster data.

All Spatial Analyst functions process the first band of any raster dataset. This section will provide an overview of raster data and how it is created.

The composition of a raster dataset

A raster dataset, like a map, describes the location and characteristics of an area and their relative positions in space. Because a single raster typically represents a single theme, such as landuse, soils, roads, streams, or elevation, multiple raster datasets should be produced to fully depict an area.

The cell

A raster dataset is made up of cells. Each *cell*, or pixel, is a square that represents a specific portion of an area. All cells in a raster must be the same size. The cells in a raster dataset can be any size that you desire, but they should be small enough to accomplish the most detailed analysis. A cell can represent a square kilometer, a square meter, or even a square centimeter.

Cell

Rows and columns

Cells are arranged in rows and columns, an arrangement that produces a Cartesian matrix. The rows of the matrix are parallel to the x-axis of the Cartesian plane, and the columns to the y-axis. Each cell has a unique row and column address. All locations in a study site are covered by the matrix.

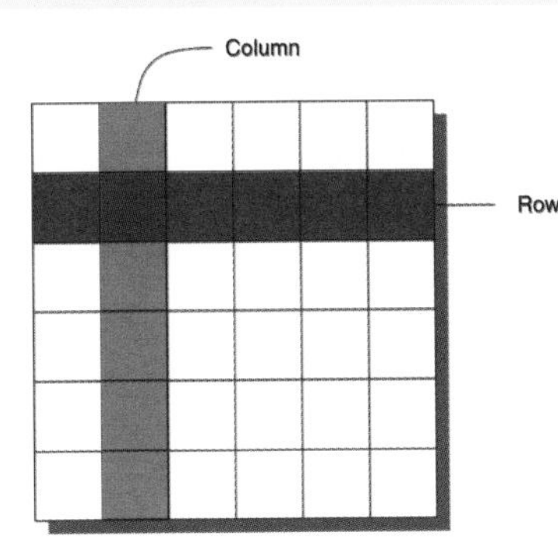

Values

Each cell is assigned a specific value to identify or describe the class, category, or group the cell belongs to or the magnitude or quantity of the phenomenon the raster describes. The characteristics the values represent include soil type, soil texture, landuse class, water body type, road class, and housing type.

A value can also represent the magnitude, distance, or relationship of the cell on a continuous surface. Elevation, slope, aspect, noise pollution from an airport, and pH concentration from a bog are all examples of continuous surfaces.

For rasters representing images and photographs, the values can represent colors or spectral reflectance.

Both *integer* and *floating-point* values are supported in Spatial Analyst. Integer values are best used to represent *categorical* data, and floating-point values to represent *continuous* surfaces.

Zones

Any two or more cells with the same value belong to the same *zone*. A zone can consist of cells that are connected, disconnected, or both. Zones whose cells are connected usually represent single features of an area, such as a building, a lake, a road, or a power line. Assemblages of entities, such as forest stands in a state, soil types in a county, or the single-family houses in a town, are features of an area that will most likely be represented by zones made up of many disconnected groups of connected cells.

Every cell on a raster belongs to a zone. Some raster datasets contain only a few zones, while others contain many.

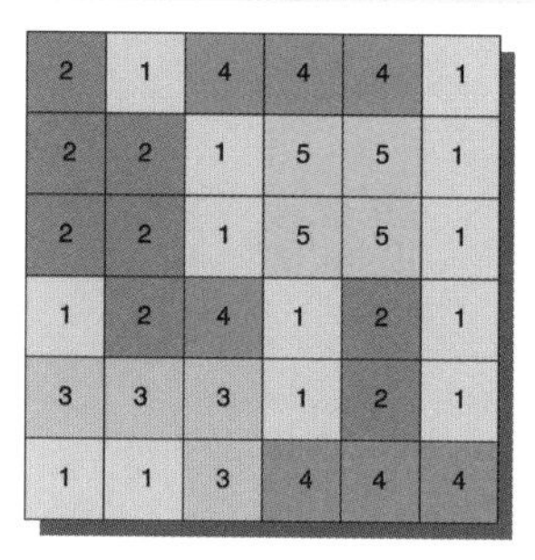

Zone with value 1
Zone with value 2
Zone with value 3
Zone with value 4
Zone with value 5

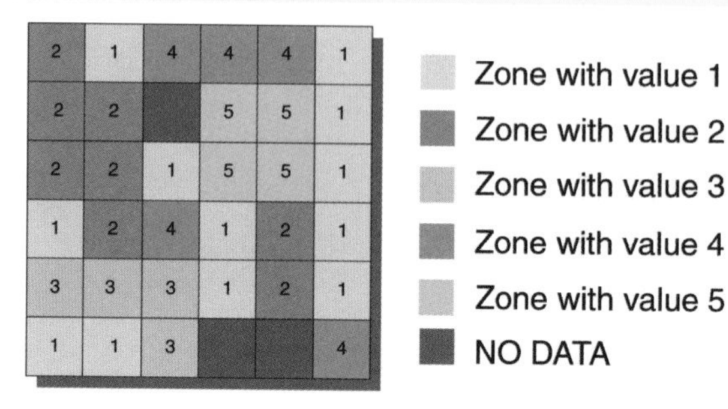

Regions

Each group of connected cells in a zone is considered a *region*. A zone that consists of a single group of connected cells has only one region. Zones can be composed of as many regions as are necessary to represent a feature; the number of cells that make up a region has no practical limits. Spatial Analyst provides the tools needed to turn regions into individual zones. In the raster dataset above, Zone 2 consists of two regions, Zone 4 of three regions, and Zone 5 of only one region.

NoData

If a cell is assigned the NoData value, then either no information or insufficient information about the particular characteristics of the location the cell represents is available. The NoData value, sometimes also referred to as the null value, is treated differently from any other value by all operators and functions.

Cells with NoData values are processed in one of two ways:

1. Assigning NoData to the output cell location if the NoData value exists for the location on any of the inputs in an operator or local function, in its neighborhood in a focal function, or in its zone in a zonal function.

2. Ignoring the NoData cell and completing the calculations with all valid values.

The second option, to ignore the NoData cell, is not possible when using operators between two datasets or with local functions. When a NoData cell is within the neighborhood of a cell in a focal function or a zone of a zonal function, by default, the sum, median, variety, majority, or minority of all cells with known values can be calculated and assigned to the output raster dataset (this default can be overridden).

The associated table

Integer—categorical—raster datasets usually have an *attribute table* associated with them. The first item in the table is Value, which stores the value assigned to each zone of a raster. A second item, Count, stores the total number of cells in the dataset that belong to each zone. Both Value and Count are mandatory items.

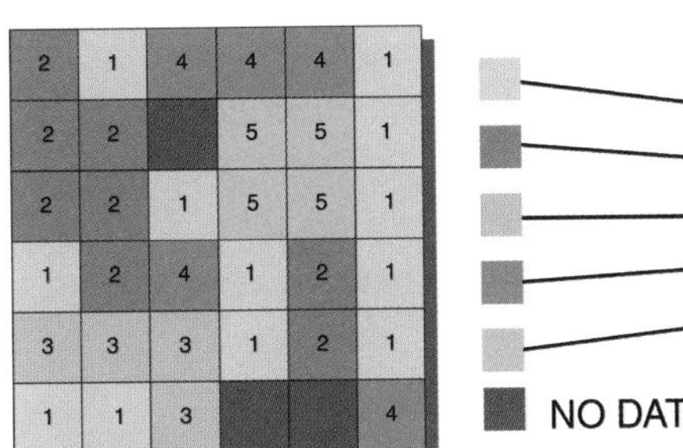

VAT	
VALUE	COUNT
1	12
2	8
3	4
4	5
5	4

An essentially limitless number of optional items can be incorporated into the table to represent the other attributes of the zone.

NO DATA

VAT				
VALUE	COUNT	TYPE	CANOPY	BUG-DAM
1	12	Maple	30	8
2	8	Oak	65	10
3	4	Field	0	0
4	5	Hickory	45	20
5	4	Pine	80	35

Name

Each raster dataset must have a name to distinguish it from the other raster datasets in a database. All access to a raster dataset is performed through its name, which must be used consistently in all expressions.

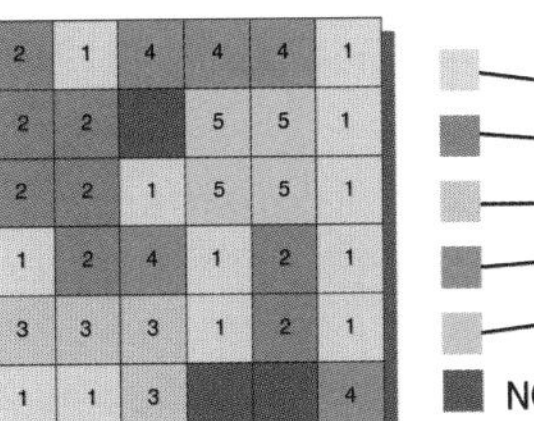

VEGETATION VAT				
VALUE	COUNT	TYPE	CANOPY	BUG-DAM
1	12	Maple	30	8
2	8	Oak	65	10
3	4	Field	0	0
4	5	Hickory	45	20
5	4	Pine	80	35

Coordinate space and the raster dataset

Coordinate space defines the spatial relationship between the locations in a raster dataset. All raster datasets are in some coordinate space. This coordinate space may be a real-world coordinate system or image space. Since almost all raster datasets represent some real-world location, it is best to have that dataset in the real-world coordinate system that best represents it. Converting a raster dataset from a nonreal-world coordinate system—image space—to a real-world coordinate system is called georeferencing.

For a raster dataset, the orientation of the cells is determined by the x- and y-axes of the coordinate system. Cell boundaries are parallel to the x- and y-axis, and the cells are square in map coordinates. Cells are always referenced by an x,y location in map coordinate space and never by specifying a row–column location.

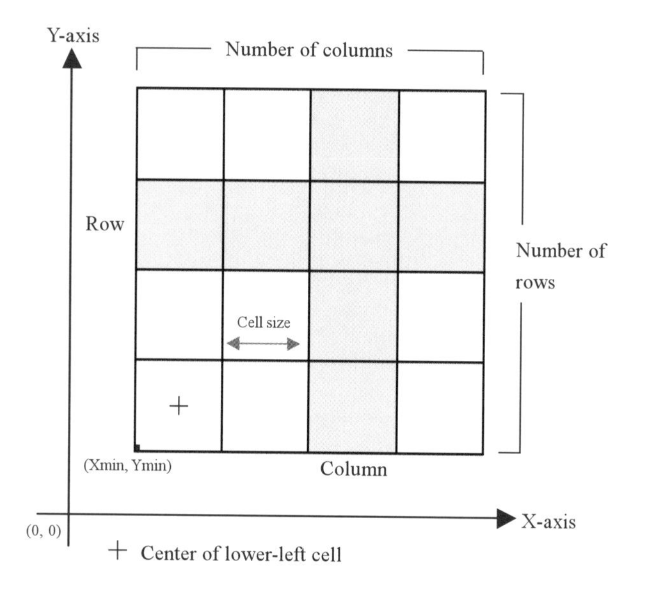

The x,y Cartesian coordinate system associated with a raster dataset that is in a real-world coordinate space is defined with respect to a *map projection*. Map projections transform the three-dimensional surface of the earth to allow the raster to be displayed and stored as a two-dimensional map.

The process of rectifying a raster dataset to map coordinates or converting a raster dataset from one projection to another is referred to as geometric transformation.

Georeferencing a raster dataset

To georeference a raster dataset from image space to a real-world coordinate system, you need to know the location of recognizable features in both coordinate spaces. These locations are used to create control points. The control points are used to build a polynomial transformation that will warp the image from one coordinate space to another. This can be done with the Georeferencing toolbar (click View, point to Toolbars, and click Georeferencing).

Control points are locations that can be accurately identified on the raster dataset and in real-world coordinates. These identifiable locations may be road and stream intersections, building corners, bridges, the mouth of a stream, rock outcrops, and identifiable points on geometric landscape features such as the end of a jetty of land, the corner of an established field, or the intersection of two hedgerows.

For each control point selected on the input raster dataset, the output location may be specified either by graphically selecting a point that is already in the desired output coordinate system or by typing in the known output coordinates. The relationship between the control points chosen in the raster dataset and the output coordinate space is then determined.

Using this relationship and a polynomial transformation, the raster dataset is converted from nonreal-world space to real-world space.

Polynomial transformation

A polynomial transformation computed using the specified control points is applied so the input locations approximate the specified output locations using a least-square fit.

The best-fit polynomial transformation yields two formulas: one for computing the output x-coordinate for an input (x,y) location and one for computing the y-coordinate for an input (x,y) location. The goal of the least-square fit is to derive a general formula that can be applied to all points, usually at the expense of slight movement of the to-positions of the control points.

When the general formula is derived and applied to the control point, a measure of the error is returned. The error is the difference between where the from-point ended up as opposed to the actual location that was specified—the to-point position. Links can be removed if the error is particularly large, and more points can be added. The more control points of equal quality used, the more accurately the polynomial can convert the input data to output coordinates.

Projecting raster datasets

The cells of a raster dataset will always be square and of equal area with respect to the Cartesian coordinate system—map coordinate space—associated with the raster dataset. The shape and area a cell represents on the surface of the earth will never be constant across a raster dataset. Since the area represented—on the face of the earth—by the cells will vary across the raster dataset, the output cell size and the number of rows and columns may change when projected.

Converting from one projection to another can also change the shape and area a cell represents on the surface of the earth. Each projection treats the relationship between a three-dimensional world and a two-dimensional one differently. You should be aware of the properties and assumptions for each projection before selecting one.

When displaying and performing analysis with raster datasets, they should be in the same coordinate space and in the same projection. If two raster datasets are in different coordinate systems, the values of the coordinates are on different scales. Errors will occur when comparing such datasets because they will represent different locations.

Geometric transformation

When you rectify a raster dataset, project it, convert the raster dataset from one projection to another, or change the cell size, you are performing a geometric transformation. Geometric transformation is the process of changing the geometry of a raster dataset from one coordinate space to another. Types of geometric transformations include rubber sheeting—usually used for georeferencing—projection—using the projection information to transform the data from one projection to another—translation—shifting all the coordinates equally—rotation—rotating all the coordinates by some angle—and changing the cell size of the dataset.

Rarely, after the geometric transformation is applied to the input raster do the cell centers of the input raster line up with the cell centers on the output raster; however, values need to be assigned to the centers.

To derive a value for the center on the output raster, a resampling technique must be used. *Resampling* is the process of determining new values for cells in an output raster that result from applying a geometric transformation to an input raster dataset. There are several techniques that can be used to derive a value. It does not matter if this transformation is a rectification, a change in projection, a change in cell size, or a rotation.

The first step in transforming a raster dataset is determining the extent of the output dataset. This is calculated by applying the transformation to the bounding box around the input raster. The output raster extent is then gridded into cells at the resolution specified for the output. If no resolution is specified, the output resolution is determined from the resolution of the input.

The coordinate value for each output cell is then determined. To find the value each cell should receive on the output raster, the center of each cell in the output must be mapped to the original input coordinate system. Each cell center coordinate is transformed backwards to identify the location of the point on the original input raster. Once the input location is identified, a value can be assigned to the output location based on the nearby cells in the input. It is rare that an output cell center will align exactly with any cell center of the input raster. Therefore, techniques have been developed to determine the output value depending on where the point falls relative to the center of cells of the input raster and the value associated with these cells. The three techniques for determining output values are nearest neighbor assignment, bilinear interpolation, and cubic convolution. Each of these techniques assigns values to the output differently, thus the values assigned to the cells of an output raster may differ according to the technique used.

Nearest neighbor assignment

Nearest neighbor assignment is the resampling technique of choice for categorical data since it does not alter the value of the input cells. Once the location of the cell's center on the output raster dataset is located on the input raster, nearest neighbor assignment determines the location of the closest cell center on the input raster and assigns the value of that cell to the cell on the output raster.

The nearest neighbor assignment does not change any of the values of cells from the input raster dataset. The value 2 in the input raster will always be the value 2 in the output raster—it will never be 2.2 or 2.3. Since the output cell values remain the same, nearest neighbor assignment should be used for nominal or ordinal data, where each value represents a class, member, or classification—categorical data such as a landuse, soil, or forest type.

Bilinear interpolation

Bilinear interpolation uses the value of the four nearest input cell centers to determine the value on the output raster. The new value for the output cell is a weighted average of these four values, adjusted to account for their distance from the center of the output cell in the input raster. This interpolation method results in a smoother-looking surface than can be obtained using nearest neighbor.

Since the values for the output cells are calculated according to the relative position and the value of the input cells, bilinear interpolation is preferred for data where the location from a known point or phenomenon determines the value assigned to the cell—that is, continuous surfaces. Elevation, slope, intensity of noise from an airport, and salinity of the groundwater near an estuary are all phenomena represented as continuous surfaces and are most appropriately resampled using bilinear interpolation.

Cubic convolution

Cubic convolution is similar to bilinear interpolation except the weighted average is calculated from the 16 nearest input cell centers and their values.

Cubic convolution will have a tendency to sharpen the data more than bilinear interpolation since more cells are involved in the calculation of the output value.

Bilinear interpolation or cubic convolution should not be used on categorical data since the categories will not be maintained in the output raster dataset. However, all three techniques can be applied to continuous data, with nearest neighbor producing a blocky output, bilinear interpolation producing smoother results, and cubic convolution producing the sharpest results.

Discrete and continuous data

Discrete data, which is sometimes called categorical or discontinuous data, mainly represents objects in both the feature and raster data storage systems. A discrete object has known and definable boundaries. It is easy to define precisely where the object begins and where it ends. A lake is a discrete object within the surrounding landscape. Where the water's edge meets the land can be definitively established. Other examples of discrete objects include buildings, roads, and parcels. Discrete objects are usually nouns.

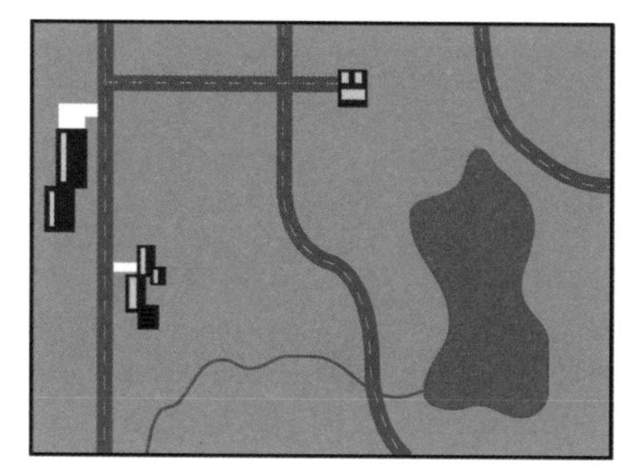

A continuous surface represents phenomena where each location on the surface is a measure of the concentration level or its relationship from a fixed point in space or from an emitting source. Continuous data is also referred to as field, nondiscrete, or surface data. One type of continuous surface is derived from those characteristics that define a surface, where each location is measured from a fixed registration point. These include elevation—the fixed point being sea level—and aspect—the fixed point being direction: north, east, south, and west.

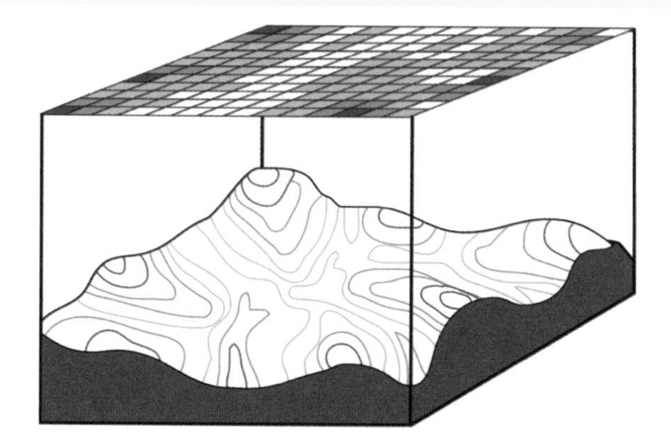

Another type of continuous surface includes phenomena that progressively vary as they move across a surface from a source. Illustrations of progressively varying continuous data are fluid and air movement. These surfaces are characterized by the type or manner in which the phenomenon moves. The first type of movement is through diffusion or any other locomotion where the phenomenon moves from areas with high concentration to those with less concentration, until the concentration level evens out. Surface characteristics of this type of movement include salt concentration moving through either the ground or water, contamination level moving away from a hazardous spill or a nuclear reactor, and heat from a forest fire. In this type of continuous surface, there has to be a source. The concentration is always greater near the source, and diminishes as a function of distance and the medium the substance is moving through.

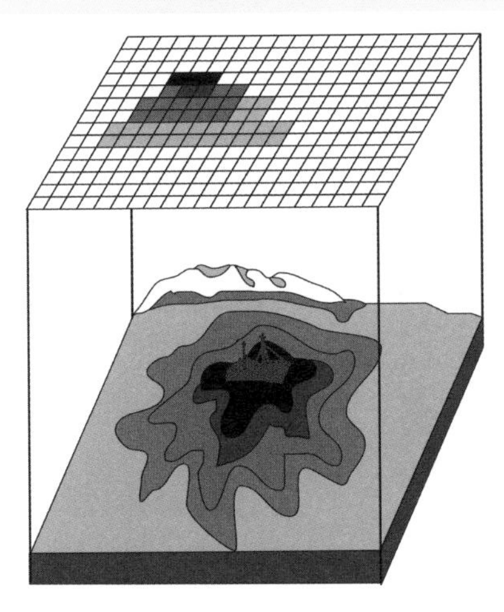

In the source-concentration surface above, the concentration of the phenomenon at any location is a function of the capability of the event to move through the medium. Another type of concentration surface is governed by the inherent characteristics of the moving phenomenon. For example, the movement of the noise from a bomb blast is governed by the inherent characteristics of noise and the medium it moves through. Mode of locomotion can also limit and directly affect the surface concentration of a feature, as is the case with seed dispersal from a plant. The means of locomotion, whether it be bees, man, wind, or water, all affect the surface concentration of seed dispersal for the plant. Other locomotion surfaces include dispersal of animal populations, potential customers of a store—car being the means of locomotion and time being the limiting factor—and the spreading of a disease.

When representing and modeling many features, the boundaries are not clearly continuous or discrete. A continuum is created in representing geographic features, with the extremes being pure discrete and pure continuous features. Most features fall somewhere between the extremes. Illustrations of features that fall along the continuum are soil types, edges of forests, boundaries of wetlands, and geographic markets influenced from a television advertising campaign.

The determining factor for where a feature falls on the continuous-to-discrete spectrum is the ease in defining the feature's boundaries. No matter where on the continuum the feature falls, the grid-cell storage can represent it to a greater or lesser accuracy.

It is important to understand the type of data you are modeling, whether it be continuous or discrete, when making decisions based on the resulting values. The exact site for a building should not be solely based on the soils map. The square area of a forest cannot be the primary factor when determining available deer habitat. A sales campaign should not be based only on the geographic market influence of a television advertising spree. The validity and accuracy of boundaries of the input data must be understood.

The resolution of a raster dataset

The size chosen for a raster cell of a study area depends on the data resolution required for the most detailed analysis. The cell must be small enough to capture the required detail, but large enough so that computer storage and analysis can be performed efficiently. The more homogeneous an area is for critical variables such as topography and landuse, the larger the cell size can be without affecting accuracy.

Before specifying the cell size, the following factors should be considered:

- The resolution of the input data
- The size of the resultant database and disk capacity
- The desired response time
- The application and analysis that is to be performed

A cell size finer than the input resolution will not produce more accurate data than the input data. It is generally accepted that the resultant raster dataset should be the same or coarser than the input data.

Spatial Analyst allows for raster datasets of different resolutions to be stored and analyzed together in the same database. Since Spatial Analyst provides this capability, the four decisions discussed above can be made separately for each dataset, rather than simultaneously for all of the rasters in the database. Raster datasets that store different types of information can be stored at different resolutions to meet the needs of the data and of the analysis that will be completed with the raster. A raster dataset representing a state's watershed boundaries can be stored at a coarser cell resolution than a raster dataset representing the distribution of endangered species.

The major disadvantage of raster-cell representation of map data is the loss of resolution that accompanies restructuring data to fixed raster-cell boundaries. Resolution increases as the size of the cell decreases; however, cost normally also increases both in disk space and processing speeds. For a given area, changing cells to half the current size requires as much as four times the storage space, depending on the type of data and the storage techniques used. For most users, the efficiency of cell-based analysis more than compensates for the loss of resolution.

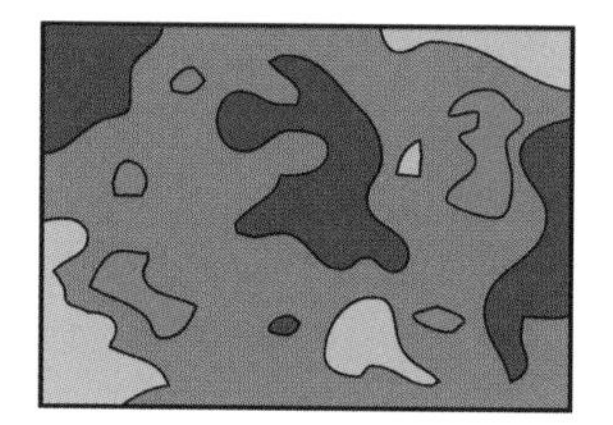

Input vegetation

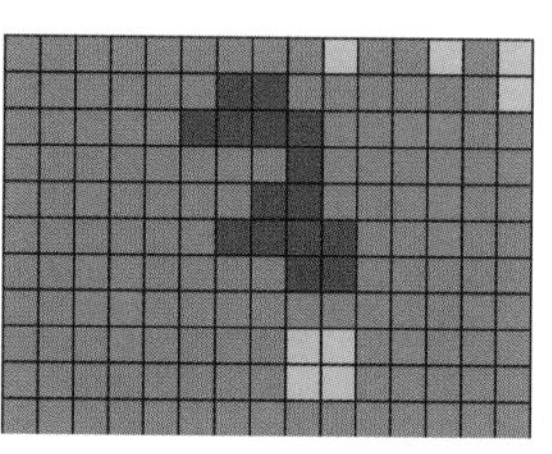

Coarse resolution

Larger cells may encompass more than one data value, which must be aggregated or prioritized and each cell given a single value, thereby decreasing data resolution. The optimum cell size to capture the appropriate detail varies from study to study. The smaller the cells, the greater the resolution and accuracy; but coding, database storage, and processing speed for analysis are more costly.

Polygons

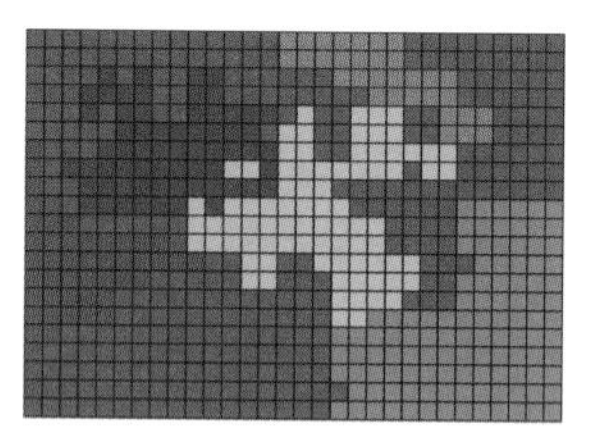

Raster from polygons

Raster encoding

The process of creating a raster dataset is like draping a fishnet containing square cells over the study area. A code is assigned to each cell according to the feature that is at the center of the cell. The code or value of a cell is a numeric value that corresponds to an attribute type. Numeric values speed processing and allow for data compression.

Each cell represents a specified portion of the world. A cell can be any size you define; there are no practical limits. The main consideration is that the size be appropriate for the analysis. For example, you would not normally use a one-kilometer cell size when studying a field mouse habitat.

If the input data is polygonal, then each cell on the resulting raster dataset from the conversion process is assigned the value of the feature that passes through the center of the cell. It is only guaranteed that the feature that the value represents is present at the center of the cell. For continuous data (see 'Discrete and continuous data' earlier in this chapter) this is the only situation that is possible. However, for discrete data, it is assumed that the feature homogeneously fills the entire extent of the cell. There is some chance that the cell center is not representative of the entire cell, but the size of the cells can be reduced if desired.

If the input features are points, then any cell extent that encompasses a point will receive the value of the attribute of the point data that is being converted. By definition, a point has no area and you are converting the data to a locational representation that has area. With a cell representation there is some generalization of the original data. If two or more points fall within the extent of a cell, Spatial Analyst randomly selects one of the points when assigning a value to the cell. Thus, it is possible to have fewer cells with values than there are points being converted. You should make your cell size small enough to capture enough of the input points for the desired analysis.

Converting linear features to a raster dataset is similar to converting point features. For any line that passes within the extent of a cell, the cell will receive the value of the attribute identified in the conversion. If multiple lines pass through a single cell, Spatial Analyst will randomly select one of the lines to represent that cell location on the output raster dataset. As with point data, linear features will be as wide as the size of the cell. For example, if the linear features that are being converted represent roads, and if the size of the cells is a kilometer, the road will be a kilometer wide on the output raster dataset. Obviously, a road is not a kilometer wide, thus you should select a cell size that is appropriate to the linear feature that you are representing. If the cell size is a meter, then the road would only be a meter wide.

For additional information and understanding of the issues of the raster encoding of the different feature types, refer to the next section, 'Representing features in a raster dataset'.

Representing features in a raster dataset

When converting points, polylines, and polygons to a raster, you should be aware of how the raster dataset will represent the features.

Point data

A point feature is any object at a given resolution that can be identified as being without area. Although a well, a telephone pole, or the location of an endangered plant are all features that can be rendered as points at some resolutions, at other resolutions they do in fact have area. For example, a telephone pole viewed from an airplane two kilometers high will be represented by only a point, but the same pole viewed from an airplane 25 meters high will be represented by a circle.

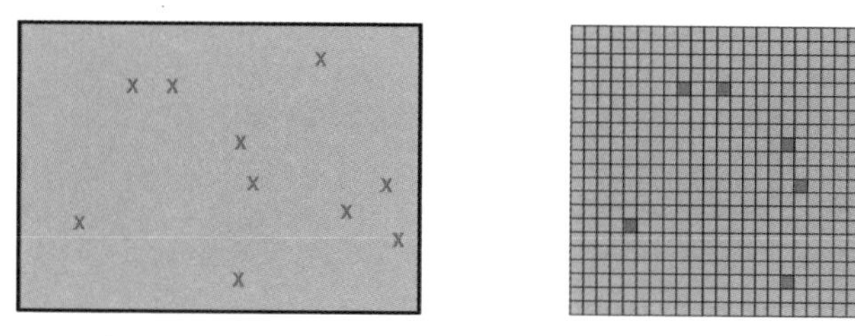

Point features *Raster point features*

Point features are represented by the smallest unit of a raster, a cell. It is important to remember that a cell has area as a property. The smaller the cell size, the smaller the area and thus the closer the representation of the point feature. Points with area have an accuracy of plus or minus half the cell size. This is the trade-off that must be made when working with a cell-based system. Having all data types—points, polylines, and polygons—in the same format and being able to use them interchangeably in the same language are more important to many users than a loss of accuracy.

Linear data

Linear data is all of those features that, at a certain resolution, appear only as a polyline, such as a road, a stream, or a power line. A line by definition does not have area. In Spatial Analyst, a polyline can be represented only by a series of connected cells. As with a point, the accuracy of the representation will vary according to the scale of the data and the resolution of the raster dataset.

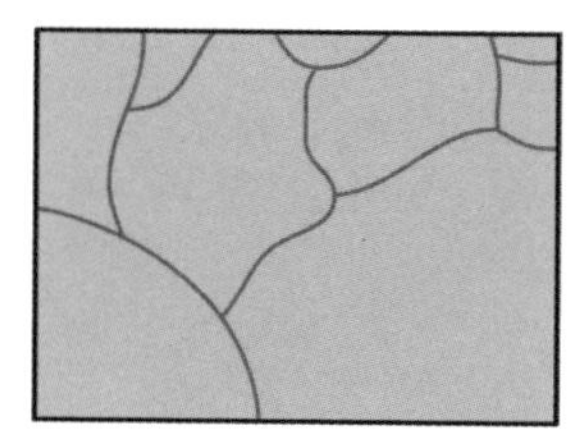

Polyline features

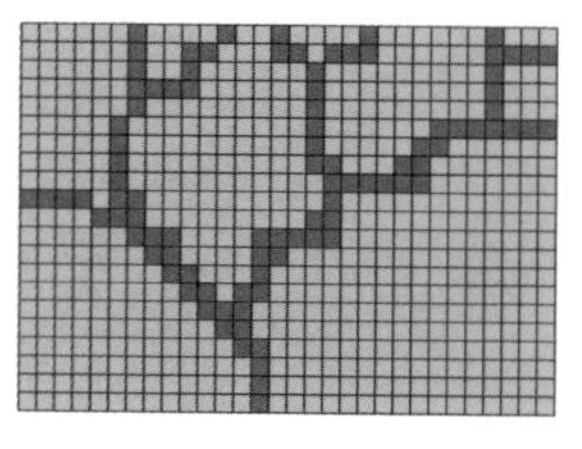

Raster line features

Polygon data

Polygonal or areal data is best represented by a series of connected cells that best portrays its shape. Polygonal features include buildings, ponds, soils, forests, swamps, and fields.

Trying to represent the smooth boundaries of a polygon with a series of square cells does present some problems, the most infamous of which is called the jaggies, an effect that resembles stair steps. Since Spatial Analyst can handle very large raster datasets with millions of cells, the jaggies become insignificant.

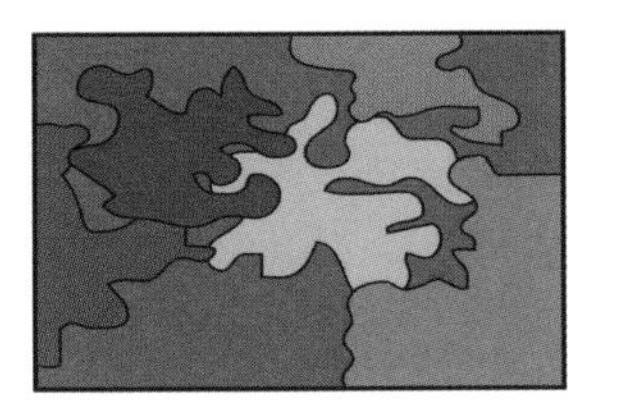

Polygon features

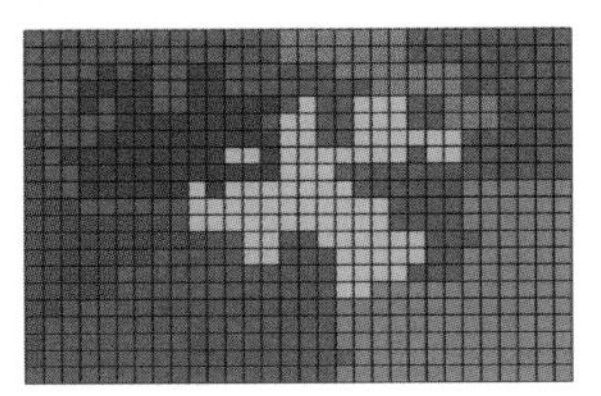

Raster polygon features

To iterate, the accuracy of the above representation is dependent on the scale of the data and the size of the cell. The finer the cell resolution and the greater the number of cells that represent small areas, the more accurate the representation.

Assigning attributes to a raster dataset

The value associated with a cell is an identifier that defines to which class, group, category, or member the cell belongs. The value is a number, either an integer or floating point. Cell locations with the same value belong to the same zone. Cells of the same zone do not have to be connected. When an integer value is used, it is often a code for a much more complex identification. For example, a 4 may equate to single-family residential parcels on a landuse raster dataset. Associated with the 4 might be a series of attributes, such as the average commercial value, average number of inhabitants, or census code. These additional attributes are either managed by the user manually or in a relational database.

There is usually a one-to-many relationship between the cell values, or codes, and the number of cells that are assigned the code. That is, there might be 400 cells with the value 4—single-family residential—and 150 cells associated with 5—commercial zoning—on the landuse raster dataset. The code is stored many times, once for each of the cells in the category—which will vary according to the storage technique—but the attributes of the code are only stored once. This reduces storage and simplifies updating.

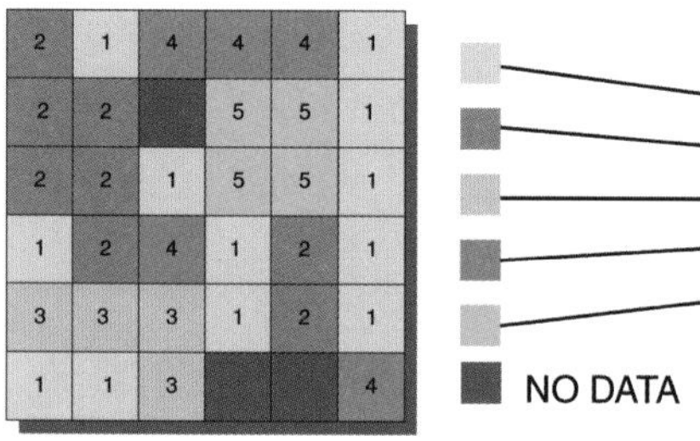

2	1	4	4	4	1
2	2		5	5	1
2	2	1	5	5	1
1	2	4	1	2	1
3	3	3	1	2	1
1	1	3			4

NO DATA

VAT		
VALUE	COUNT	TYPE
1	12	Maple
2	8	Oak
3	4	Field
4	5	Hickory
5	4	Pine

The field you use in the conversion process will affect the analysis that you can perform on the dataset. If you have a polygon feature dataset that contains landuse type and owner for each parcel in a town, you can use either attribute. If you use landuse type you will be able to ask questions, such as "Where are all the agricultural areas that are available for building?" However, you cannot add (join) the owner attribute to the raster dataset because this is a many-to-one relationship. That is, there will probably be many owners for parcels containing forest. If you use owner, you can then ask questions, such as "Which parcels does Fred Smith own?" You can also relate the landuse type to the relational table because each parcel will have one landuse type. Where this logic collapses is when one person owns multiple parcels with different landuse types. In this case, you may have to use parcel-ID or some other unique feature when converting.

Usually with continuous data, each cell has a unique value and will have no attribute table, so there will be no attributes to relate. In this case, the many-to-one issue will not be applicable.

When creating a raster dataset, the value and level of groupings should reflect the analysis to be completed. Whether slope is divided into five categories—0 to 10, 11 to 20, 21 to 30, 31 to 40, and 50+ percent—or divided into groupings containing only two percent intervals—0 to 2 percent, 3 to 4 percent, and so forth—depends on the most detailed breakdown necessary in future analyses. When in doubt, the most detailed breakdown should be used. Information can be grouped into fewer categories more easily than splitting fewer categories into more categories.

Using feature data directly in Spatial Analyst

Several of the Spatial Analyst dialog boxes allow you to enter a point, polyline, or polygon feature directly into the function. There are two ways that features are handled in Spatial Analyst. It either processes the feature data directly, or it converts it to a raster and then processes it.

Certain functions require that one or more of the inputs be feature data, and Spatial Analyst processes the data as feature data. For instance, the Inverse Distance Weighted and the Kriging functions create a continuous surface raster dataset from a point feature layer of measured sample data. The calculations are performed on the point feature data directly.

Other functions allow you to enter feature data for one or more of the inputs and converts that feature data to a raster before performing the calculations. An example of such a function is the zone dataset in the zonal statistics function. Note that the value raster must be a raster dataset. Feature data entered for the zone dataset will take the cell size specified on the cell size tab of the Options dialog box when it is converted to a raster. The output resolution can be a specific cell size or the maximum or minimum cell size of the other input raster datasets into the function. The default is set to the coarsest input raster dataset into the function. For additional information on issues related to converting feature data, please refer to 'Representing features in a raster dataset' earlier in this chapter.

You will know if the input can be either a feature or a raster dataset because when you open the browser to enter the input, the browser will say "Raster datasets and feature classes" in the the Show of type input field, and both feature and raster data will be displayed in the browser.

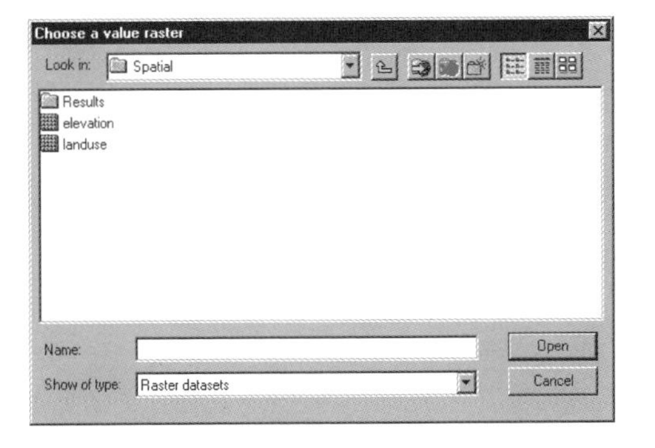

If only rasters are allowed as input, then the browser will say "Raster Datasets" in the Show of type input field, and only raster datasets will be displayed.

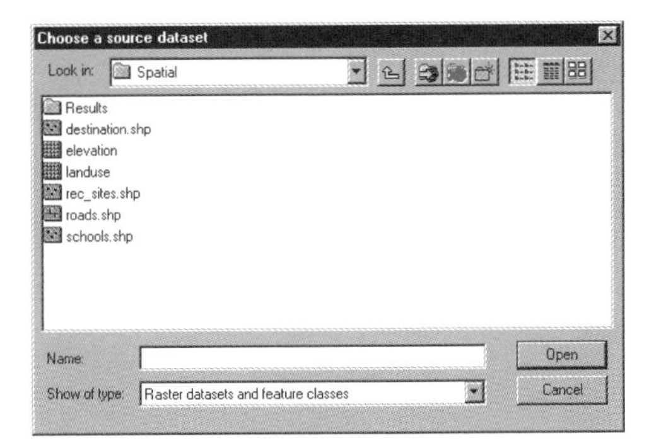

Some browsers will allow the input of both feature and raster data. Because of this capability, you do not have to convert all the feature data to raster datasets before you perform analysis.

Deriving raster datasets from existing maps

When creating raster datasets from existing maps—data entry—everal factors must be considered to allow for full utilization of the input data.

Selecting maps

When selecting maps for the creation of the database, you must be aware of:

- The age and date of the map
- The cartographic accuracy
- The resolution and detail
- The compatibility of the map with other input maps

The age and date will determine whether the map is current enough for the analysis to be completed. If it is not, then a more current map must be found. The cartographic accuracy, resolution, and detail must be fine enough to complete the analysis at hand, but not so detailed that the cost of entry, loss in processing speed, and size of the database are too inhibiting. The input maps must be compatible. Simple things, such as the location of registration points on different maps, can ease database construction. Accuracy of analysis depends on the consistency and accuracy of the data variables.

Potential errors

Even with current maps that are accurate, at the same resolution, with the desired amount of detail, and compatible, errors can still occur. Some of the most common errors include:

- Drafting errors
- Different cartographic projections used to draft the original data
- Different photographic projections used to draft the original data
- Physical changes in the materials used for the maps—shrinking or swelling

Drafting errors can be minimized by taking care in drawing, allotting sufficient time for drawing and entry, rotating staff frequently from the mundane tasks, and assigning dependable individuals to the tasks. Maps in different projection systems can be registered in the computer, but you must remember to transform these layers to the desired projection at a future time. It is difficult to monitor the photographic projections, but it can be requested from the supplier that the images be in the same projections before drafting or that the data be transformed in the computer if necessary. Using Mylar® in a temperature-controlled environment will eliminate shrinking and swelling problems.

Understanding cell-based modeling 5

IN THIS CHAPTER

- **Understanding analysis in Spatial Analyst**
- **The operators and functions of Spatial Analyst**
- **NoData and how it affects analysis**
- **Values and what they represent**
- **The analysis environment**
- **The cell size and analysis**
- **Handling projections during analysis**

One of the strongest aspects of Spatial Analyst is its analytical capabilities. Spatial Analyst takes a locational perspective, where each cell represents a location and the value associated with each cell identifies the type of phenomenon that is at each location (see Chapter 4, 'Understanding raster data'). Each operator and function in Spatial Analyst manipulates the value for each cell in different ways, depending on the type of function.

In this chapter you will learn the general principles of cell-based modeling. By combining these principles you will be able to solve almost any of your specific problems. Not only will you learn about the general principles of cell-based modeling, but you will also learn what considerations you must be aware of when performing analysis. You will understand the effects that the values in the raster dataset, cell size, NoData, projections, and analysis extent will have on your analysis. It is from this understanding that you will make better decisions when performing cell-based analysis.

Understanding analysis in Spatial Analyst

The easiest way to understand cell-based modeling is from the perspective of an individual cell—the worm's-eye approach—as opposed to the entire raster—the bird's-eye approach. To do so, think of yourself as a cell in a raster dataset. You represent a location and you have a value. All Spatial Analyst operators and functions will ask you to manipulate your value—or remain the same—based on a set series of rules.

For you to calculate an output value for your location using any Spatial Analyst operation or function, there are three things you need to know.

- You need to know your value.
- You need to know the manipulation of the operator or function.
- You need to know which other cell locations and their values need to be included in your calculations.

How do you determine these three things?

You automatically know what your value is for your location.

Each operator and function in Spatial Analyst manipulates the value at your location in different ways. You will know how to manipulate your value based on the operator or function being applied from knowledge built into Spatial Analyst.

With some Spatial Analyst operations and functions you can calculate an output value just from knowing the value of your location, such as raising your value by a specified power [a local function]. To complete other operations and functions, you need to know the values of other locations within the raster dataset you belong to, such as looking in a neighborhood around you [a focal function], or you will include cell locations and their values belonging to other raster datasets, such as zonal functions.

Let's walk through the three-step process for several functions. If the Cos function is being applied to your raster dataset, for you to return an output value for your location, you will need to know your value and how to take the cosine of your value. If a focal function is applied, looking for the maximum value within a 3-by-3 neighborhood (you will learn more about focal functions later in this chapter), you must know your value and the values of the eight immediate neighbors around you. You will assign the highest of the nine values to your location on the output raster dataset. If a zonal function is applied with the mean option (you will learn more about the zonal functions later in this chapter), you need to know your value and must take the mean of all the values of the cell locations that belong to the same zone as you, defined by a zone raster dataset. If the addition operator is applied to your raster dataset and to two other raster datasets, you need to add your value to the values of the same location you represent in the two other raster datasets in order to return an output value for your location. If the Straight Line—Euclidean—Distance function is applied, you will need to determine how far you are from the closest source—which is defined by a source dataset—in order to return an output value for your location.

The three-step process occurs for each location in the raster dataset. All operators and functions work on a cell-by-cell basis, and each calculation for each cell needs to know the value of the cell, the manipulation that is being applied, and which other cell locations to include in the calculations. The Spatial Analyst operators and functions are grouped into categories based on how they manipulate values. Instead of trying to memorize each operator and function, you need only to understand how the cell values are manipulated in the different categories.

For many functions you can refine how the manipulation—the calculations—will be performed through parameters. For example, for a focal function, the cells to include may vary based on the neighborhood that's specified.

The operators and functions of Spatial Analyst

The functions associated with raster-cell cartographic modeling can be divided into five types:

- Those that work on single cells (*local functions*)
- Those that work on cells within a neighborhood (*focal functions*)
- Those that work on cells within zones (*zonal functions*)
- Those that work on all cells within the raster (*global functions*)
- Those that, when combined in a series, perform a specific application (*application functions*)

Each of these categories is influenced by, or is based on, the spatial or geometric representation of the data, not solely on the attributes that they portray. That is, a function that is to add two layers together—work on single cells—is dependent on the location and the value of its counterpart on another layer. Functions applied to cells within neighborhoods or zones rely on the spatial configuration of the neighborhood or zone as well as the cells and values in the configuration.

Local functions

Local, or per-cell, functions compute an output raster dataset where the output value at each location is a function of the value associated with that location on one or more raster datasets. That is, the value of the single cell, regardless of the values of neighboring cells, has a direct influence on the value of the output. A per-cell—local—function can be applied to a single raster dataset or to multiple raster datasets. For a single dataset, examples of per-cell functions are the trigonometric functions—for example, sin—or the exponential and logarithmic functions—for example, exponential or log.

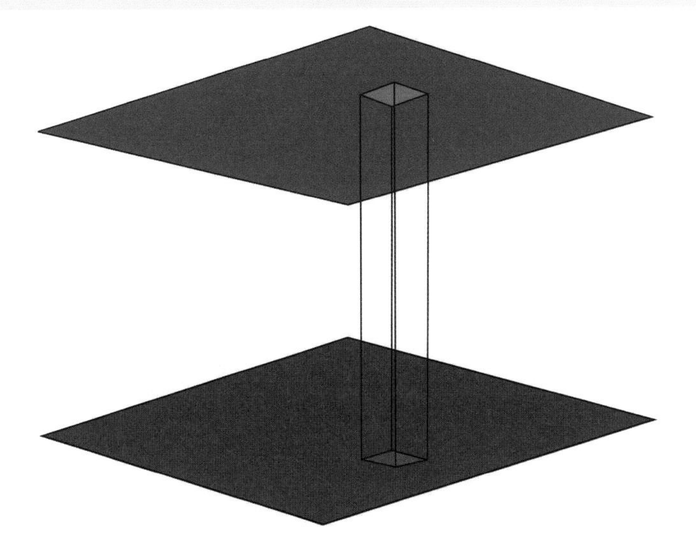

Examples of local functions that work on multiple raster datasets are functions that return the minimum, maximum, majority, or minority value for all the values of the input raster datasets at each cell location.

Focal functions

Focal, or neighborhood, functions produce an output raster dataset in which the output value at each location is a function of the input value at a location and the values of the cells in a specified neighborhood around that location. A neighborhood configuration determines which cells surrounding the processing cell should be used in the calculation of each output value.

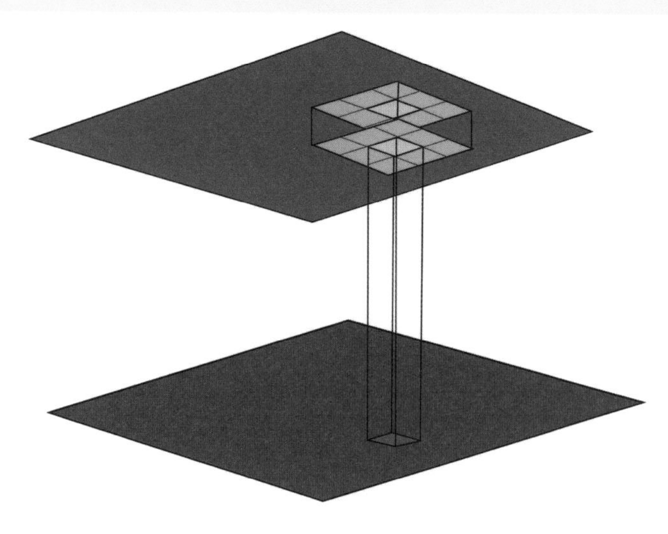

Neighborhood functions can return the mean, standard deviation, sum, or range of values within the immediate or extended neighborhood.

Zonal functions

Zonal functions compute an output raster dataset where the output value for each location depends on the value of the cell at the location and the association that location has within a cartographic zone. Zonal functions are similar to focal functions except that the definition of which cells to include in the processing—the neighborhood—in a zonal function is defined by the configuration of the zones or features in the input zone dataset, not by a specified neighborhood shape. Each zone can be unique. Operations that can be completed on these cells return the mean, sum, minimum, maximum, or range of values from the first dataset that fall within a specified zone of the second.

Global functions

Global, or per-raster, functions compute an output raster dataset in which the output value at each cell location is potentially a function of all the cells in the input raster datasets. There are two groups of global functions: Euclidean distance and weighted distance.

Euclidean distance global functions assign to each cell in the output raster dataset its distance from the closest source cell—a source may be the location from which to start a new road. The direction of the closest source cell can also be assigned as the value of each cell location in an additional output raster dataset.

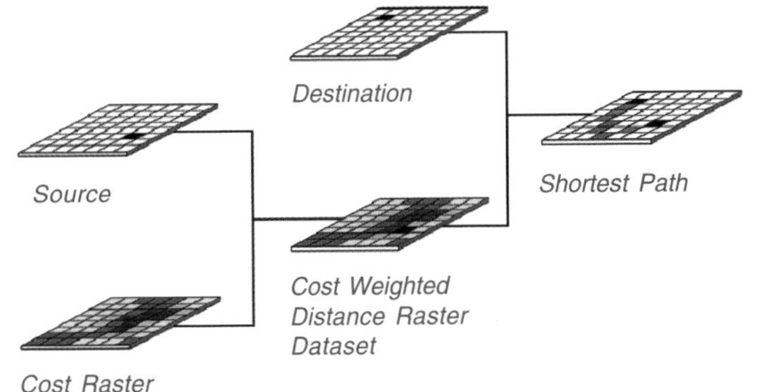

By applying a global function to a weighted—cost—surface, you can determine the cost of moving from a destination cell—the location where you wish to end the road—to the nearest source cell. To take this one step further, the shortest path over a cost surface can be calculated over a nonnetworked surface from a source cell to a destination cell. In all the global calculations, knowledge of the entire surface is necessary to return the solution.

Application functions

There is a wide series of cell-based modeling functions that are developed to solve specific applications. The local, focal, zonal, and global functions are not specific to any application. There is some overlap in the categorization of an application function and the local, focal, zonal, and global functions—such as the fact that even though slope is usually used in the application of analyzing surfaces, it is also a focal function. Some of the application functions are more general in scope, such as surface analysis, while other application functions are more narrowly defined, such as the hydrologic analysis functions. The categorization of the application functions is an aid to group and understand the wide variety of Spatial Analyst operators and functions. You may find that a specific application function can manipulate cell-based data for an entirely different application than its category.

Some of these application functions are available in the Spatial Analyst user interface. Others are available in other dialog boxes such as the Georeferencing toolbar. Some are available through sample applications built by ESRI and other users, while others are only available through the Raster Calculator via Map Algebra (see Appendix A) or through the Spatial Analyst object model.

The following sections provide an overview of the application functions. For additional information on the application functions that are available through the Spatial Analyst user interface, see Chapter 7, 'Performing spatial analysis', and refer to the online command references for information on those application functions available only through Map Algebra and the object model.

Density

The Density function distributes a measured quantity of an input point layer throughout a landscape to produce a continuous surface. For example, a retail store chain has multiple stores in a particular district. For each store, management has sales figures on its customers. Management assumes that customers patronize one store over another based on how far they have to travel. In this example, it is assumed that the customer will always choose the closest store. The farther away from the closest store, the farther the customer will need to travel to that store. But shoppers farther away will also shop at other stores. Management wishes to study the distribution of where the customers live. From the sales figures and the spatial distribution of the stores, management wishes to create a surface of their customers by intelligently spreading them out across the landscape. To accomplish this task, Spatial Analyst considers where each store is in relation to other stores, the quantity of customers shopping at each store, and how many cells need to share a portion of the measured quantity, the shoppers. The cells nearer to the measured points, the stores, receive higher proportions of the measured quantity than those farther away.

Surface generation

The surface functions use the surface representation of a raster dataset to represent height, concentration, or magnitude—for example, elevation, pollution, or noise.

Surface generation functions, called surface interpolators, create a continuous surface from sampled point values. Surface generation functions make predictions for all locations in a raster dataset whether a measurement has been taken at the location or not. There are a variety of ways to derive a prediction for each location; each method is referred to as a model. With each model, there are different assumptions made of the data—for example, the data needs to be normally distributed)—and the model produces predictions using different calculations. Below is a brief description of each model that is available in Spatial Analyst.

Inverse Distance Weighted (IDW) is based on a basic principle of geography, that things close to one another are more alike. Thus, if you are at a location with no measured value, in IDW, you will look within a specified neighborhood—or distance—around you and identify the measured values. You know that things close to you are probably more like your unknown value, thus they will influence your prediction more than those measured locations farther away; you will weight the closer measured points more than those farther away. Hence, the name IDW—as the distance increases, you will inversely weight the values. This process continues for each location in the study site.

Polynomial trend surface is conceptually similar to taking a piece of paper and trying to pass it through measured points that are raised to the height of their values. That paper is fitted so that overall it fits best to all the points.

Spline is conceptually like taking a rubber membrane and, once the measured points are raised to the height of their values, trying to fit it through the points the best you can. The criterion imposed on fitting this membrane is that it must pass through the measured points.

Kriging is a statistical method that quantifies the correlation of the measured points through variography. When making a prediction for an unknown location, kriging weights the nearby measured points by their configuration around the prediction location and uses the fitted model from variography to determine a value. For additional information on kriging, see Chapter 7, 'Performing spatial analysis'.

ArcGIS Geostatistical Analyst provides additional tools for more advanced surface generation.

Surface analysis

The premise behind the surface analysis functions is that additional information can be derived by producing new data and identifying patterns in existing surfaces.

Slope identifies the slope, or maximum rate of change, from each cell to its neighbors. An output slope raster dataset can be calculated as either a percentage of slope—for example, 10 percent slope—or a degree of slope—for example, 45-degree slope.

Aspect identifies the steepest downslope direction from each cell to its neighbors. The value of the output raster dataset represents the compass direction of the aspect: 0 is true north, a 90-degree aspect is to the east, and so forth.

Hillshade is used to determine the hypothetical illumination of a surface for either analysis or graphical display. For analysis, hillshade can be used to determine the length of time and intensity of the sun in a given location. For graphical display, hillshade can greatly enhance the relief of a surface.

Viewshed identifies either how many of the observation points specified on the input observation raster dataset can be seen from each cell or which cell locations can be seen from each observation point.

Curvature measures the slope of the surface at each cell. It calculates the second derivative of the input-surface raster dataset—the slope of the slope. The result of the curvature function can be used to describe the physical characteristics of a surface, such as the erosion and runoff processes within a landscape. The slope identifies the overall rate of downward movement, and aspect defines the direction of flow. The profile curvature is the shape of the surface in the direction of the slope. The planform curvature defines the shape of the surface perpendicular to the direction of the slope.

Contour produces an output polyline dataset. The value of each line represents all contiguous locations with the same height, magnitude, or concentration of whatever the values on the input dataset represent. The function does not connect cell centers; it interpolates a line that represents locations with the same magnitude.

Hydrologic analysis

The shape of a surface determines how water will flow across it. The hydrologic modeling functions provide methods for describing the hydrologic characteristics of a surface. Using an elevation raster dataset as input, it is possible to model where water will flow, create watersheds and stream networks, and derive other hydrologic characteristics.

Watersheds for each section of a stream network

The hydrologic modeling functions are available through the RasterHydrologyOp or through Map Algebra via the Raster Calculator.

Geometric transformation

The geometric transformation functions either change the location of each cell in the raster dataset or alter the geometric distribution of the cells within a dataset to correct a distortion. The mosaicking functions—another geometric transformation—combine multiple raster datasets representing adjacent areas into a single raster dataset.

There are two groups of geometric transformation functions, translation and rotation, which change the location represented by the cells. Translation shifts the coordinates of the raster dataset by a specified x,y offset, and rotation rotates a raster dataset by a specified amount.

Flip and mirror are special cases of rotation functions. Using flip, a raster dataset can be flipped in a specified y direction. With mirror, a raster dataset can be mirrored in the x direction.

The geometric transformation functions that alter the geometric distribution within a raster dataset change the count of cells in some areas to correct geometric distortion. Geometric distortion occurs when features in a raster dataset are not located where they should be in the real world. From a known set of real-world coordinates that match known locations in a raster dataset, the raster cell locations can be adjusted to more closely represent reality. Warp uses a polynomial transformation to correct for distortion in the entire raster dataset.

Merge and mosaic combine several spatially adjacent raster datasets into a single, larger dataset. The difference between the two is in how they handle the overlapping areas between the raster datasets. In merge, the cell is assigned to the last input value from the series of input raster datasets. Mosaic smoothes the transition between the adjacent raster datasets in the overlapping areas. These functions are used when several raster datasets come from a tiled, continuous data source, such as adjacent satellite scenes, neighboring towns, or states that are separately managed.

Some of the geometric transformation functions are available on the Georeferencing toolbar in ArcMap, and all are available through Map Algebra, which is accessed through the Raster Calculator.

Generalization

Sometimes a raster dataset contains data that is erroneous or irrelevant to the analysis at hand or is more detailed than you need. For instance, if a raster dataset was derived from the classification of a satellite image, it may contain many small and isolated areas that are misclassified. The generalization functions assist with identifying such areas and automating the assignment of more reliable values to the cells that make up the areas.

The generalization functions are available through RasterGeneralizeOp or through Map Algebra via the Raster Calculator.

These tools provide capabilities for aggregation, edge smoothing, intelligent noise removal, and more.

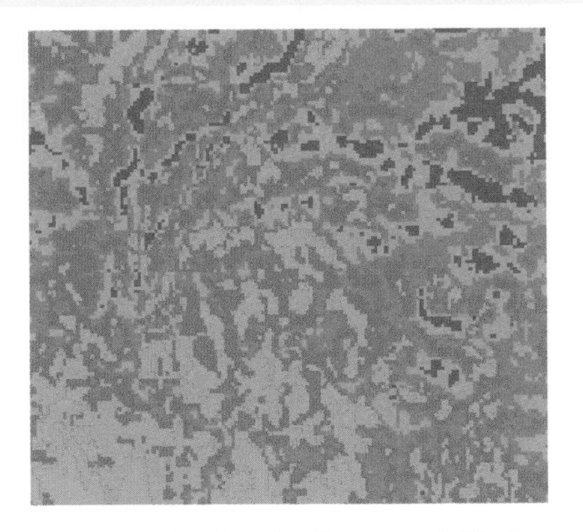

The base classification from a satellite image

Use Nibble to remove the single, misclassified cells in the classified image. This function will remove small areas of misclassified cells and assign them the most common value in their immediate neighborhood. For example, you may want to get rid of all groups of cells that are less than 7,200 square meters. These cells may be either misclassified or too small for upcoming analysis.

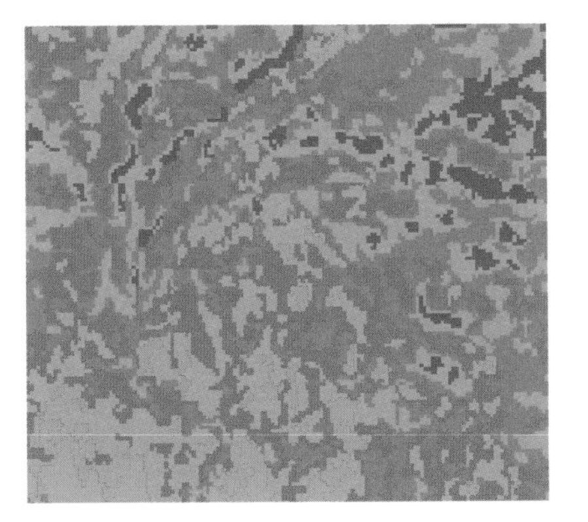

Effect of Nibble applied to the base classification

Other generalization functions include BoundaryClean and MajorityFilter, which smooth the boundaries between different zones; Expand, which expands specified zones; Shrink, which shrinks specified zones; and Thin, which thins linear features in a raster dataset and is particularly useful for cleaning up scanned paper maps.

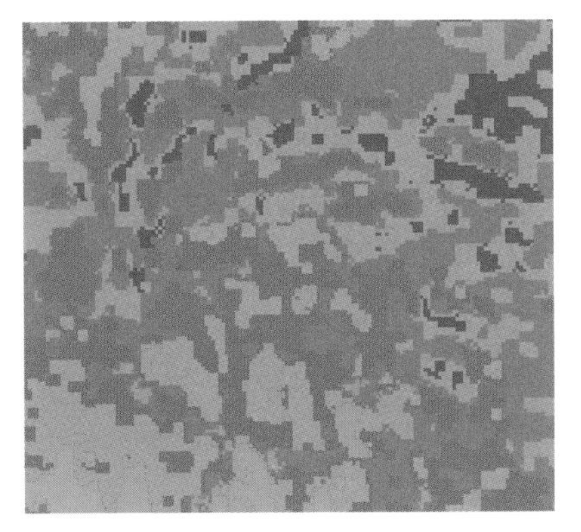

Effect of MajorityFilter applied to the output from Nibble

Resolution altering

The resolution altering functions change the resolution of an existing raster dataset. If you have one raster dataset at a finer resolution than the rest of the raster datasets, you may wish to resample the finer resolution dataset to the same resolution of the coarser ones to make all the raster datasets the same resolution. This speeds up processing and reduces the data size.

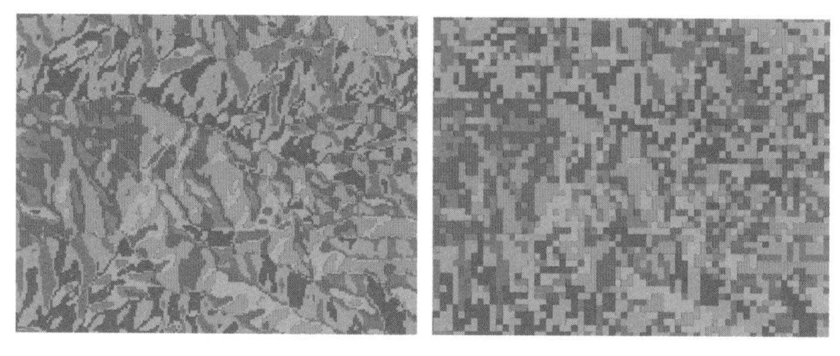

The effect on the raster of resampling to a coarser resolution

The two principal ways to determine values when changing the resolution of a raster dataset are interpolation and aggregation. One group of resampling interpolation functions uses either the nearest-neighbor, bilinear, or cubic methods on the values of the input raster dataset. A second group of resampling interpolation functions uses a specified statistical aggregation method within a neighborhood to derive values. Unlike the cell-size setting in the analysis environment, the resolution-altering functions are applied only to the resultant dataset.

The aggregation functions group a series of cells to the same value. To perform an aggregation, the block functions are implemented. With a block function, Spatial Analyst calculates a specified statistic within nonoverlapping neighborhoods.

NoData and how it affects analysis

Every cell location in a raster has a value assigned to it. When inadequate information is available for a cell location, the location can be assigned NoData. NoData and 0 are not the same; 0 is a valid value.

The fact that a location can have NoData instead of a valid value has ramifications in operators and functions. NoData means that not enough information is known about a cell location to assign it a value. There are two ways that a location with NoData can be treated in the computation of an expression:

- Return NoData for the location no matter what.
- Ignore the NoData and compute with the available values.

Depending on the operator or function, one of the above approaches will make greater sense than the other. For instance, when adding two raster datasets together, if a cell location in one of the datasets contains NoData, there is no basis for assigning a value to the corresponding location on the output raster dataset. In contrast, when looking for the minimum value in a neighborhood that contains a NoData value, an assumption can be made, or a risk taken, that the cell location with the NoData value will not be the minimum value. The focal function can thus be used to return the minimum value of the remaining valid values in the neighborhood.

Spatial Analyst fully supports the NoData concept. If NoData exists in any of the input raster datasets in the Spatial Analyst expression, the output values will be affected. The behavior of NoData is addressed for each operator and function in the online command references.

It is important to understand how NoData is handled in a particular function before making a decision. You may need to know if a location with NoData on the output ever had a value, or if it received NoData from the operator or function. Sometimes, when locations receive values, it may be important to know if the output value really is the actual minimum or maximum value, or if it is the minimum or maximum value of the existing known values.

Values and what they represent

The type of measurement system used may have a dramatic effect on the interpretation of the resulting values. A distance of 20 kilometers is twice as far as 10 kilometers, and something that weighs 100 pounds is a third as much as something that weighs 300 pounds. But someone who came in first place may not have done three times as well as someone in third place, and soil with a pH of 3 is not half as acidic as soil with a pH of 6.

To carry this even further, someone that is 60 years old is twice as old as someone that is 30 years old. But the older of the two individuals can only be twice as old as the younger individual just once in a lifetime. Also, if their birth dates are examined, and if the older individual was born in 1930 and the younger was born in 1960, the value 1930 is not twice the value 1960.

The significance of this discussion on numbers is that all numbers cannot be treated the same. It is important for you to know the type of measurement system being used in the raster dataset so that the appropriate operations and functions can be implemented and the results will be predictable. Measurement values can be broken into four types: ratio, interval, ordinal, and nominal.

Ratio

The values from the ratio measurement system are derived relative to a fixed zero point on a linear scale. Mathematical operations can be used on these values with predictable and meaningful results. Examples of ratio measurements are age, distance, weight, and volume.

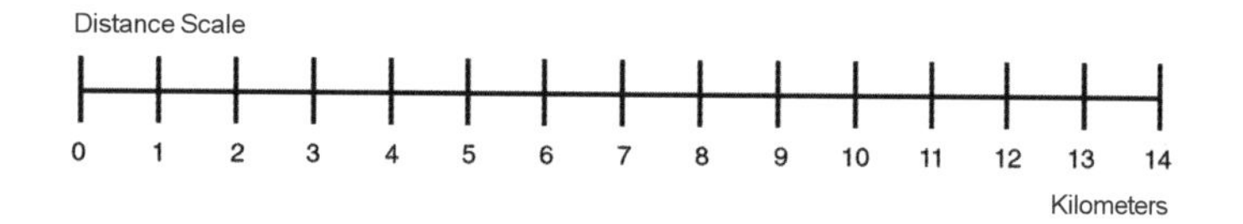

Interval

Time of day, years on a calendar, the Fahrenheit temperature scale, and pH value are all examples of interval measurements. These are values on a linear calibrated scale, but they are not relative to a true zero point in time or space. Because there is no true zero point, relative comparisons can be made between the measurements, but ratio and proportion determinations are not as useful.

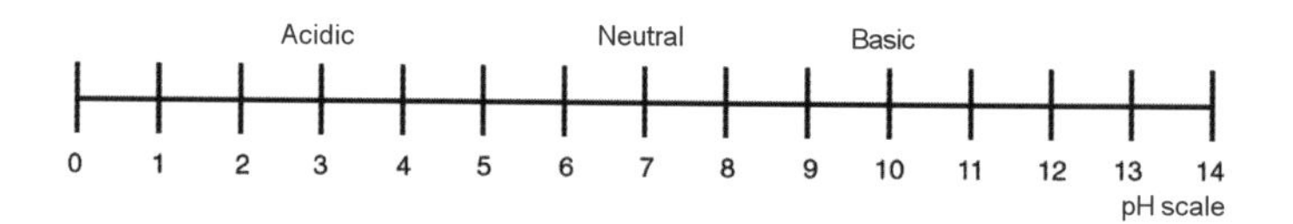

Ordinal

Ordinal values determine position. These measurements show place, such as first, second, and third, but they do not establish magnitude or relative proportions. How much better, worse, prettier, healthier, or stronger something is cannot be demonstrated from ordinal numbers.

Nominal

Values associated with this measurement system are used to identify one instance from another. They may also establish the group, class, member, or category with which the object is associated. These values are qualities, not quantities, with no relation to a fixed point or a linear scale. Coding schemes for landuse, soil types, or any other attribute qualify as a nominal measurement. Other nominal values are social security numbers, ZIP Codes, and telephone numbers.

Spatial Analyst does not distinguish between the four different types of measurements when asked to process or manipulate the values. Most mathematical operations work well on ratio values, but when interval, ordinal, or nominal values are multiplied, divided, or evaluated for the square root, the results are typically meaningless. On the other hand, subtraction, addition, and Boolean determinations can be very meaningful when used on interval and ordinal values. Attribute handling within and between raster datasets is most effective and efficient when using nominal measurements.

Discrete versus continuous data

A second subdivision of the values assigned to each cell is whether the values represent discrete or continuous data.

Discrete data

Discrete data, sometimes called categorical data, most often represents objects. These objects usually belong to a class—for example, soil type—a category—for example, landuse type—or a group—for example, political party. A categorical object has known and definable boundaries.

An integer value is normally associated with each cell in a discrete raster dataset. Most integer raster datasets can have a table that carries additional attribute information. Floating-point values can be used to represent discrete data.

Discrete data is best represented by ordinal or nominal numbers.

Continuous data

A continuous raster dataset, or surface, can be represented by a raster with floating-point values—referred to as a floating-point raster dataset—or integer values. The value for each cell in the dataset is based on a fixed point—such as sea level—a compass direction, or the distance of each location from a phenomenon in a specified measurement system—such as the noise in decibels at various sites near an airport. Examples of continuous surfaces are elevation, aspect, slope, the radiation levels from a nuclear plant, and the salt concentration from a salt marsh as it moves inland.

Floating-point raster datasets usually do not have a table associated with them because most, if not all, cell values are unique, and the nature of continuous data excludes other associated attributes.

Continuous data is best represented by ratio and interval values.

Many times, meaningless results will occur when combining discrete and continuous data, for instance, adding landuse—discrete data—to elevation—continuous data. A value of 104 on the resulting raster dataset could have been derived from adding single-family housing landuse type, with a value of 4, to an elevation of 100.

The analysis environment

Spatial Analyst allows you to process a subset of cells and to specify the resolution in which to process them. For a more detailed discussion of the Spatial Analyst analysis environment, see Chapter 6, 'Setting up your analysis environment'.

The analysis extent

When performing analysis, the area of interest may be a portion of a larger raster dataset. If the area of interest is a portion of a larger raster dataset, the analysis extent can be set to encompass only the desired cells. All subsequent results from analysis will be to this extent. The analysis extent is a rectangle and is specified by identifying the coordinates of the window in map space.

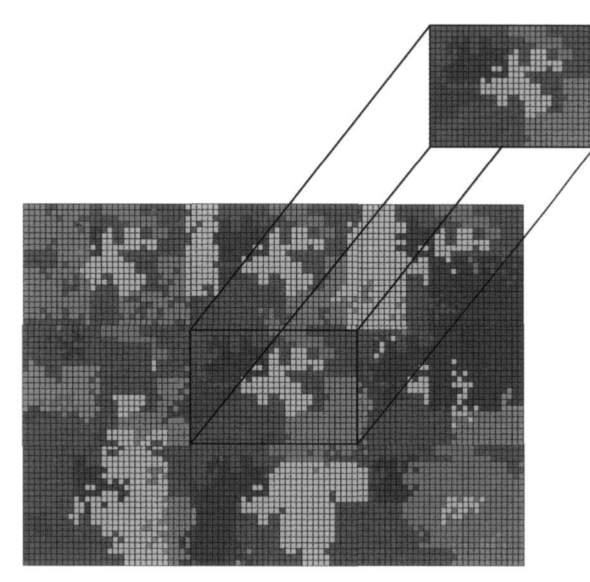

Performing analysis on a small section of the raster

The mask

The mask identifies those cells within the analysis extent that will not be considered when performing an operation or a function. All identified cells will be masked out and assigned to the NoData value on all subsequent output raster datasets.

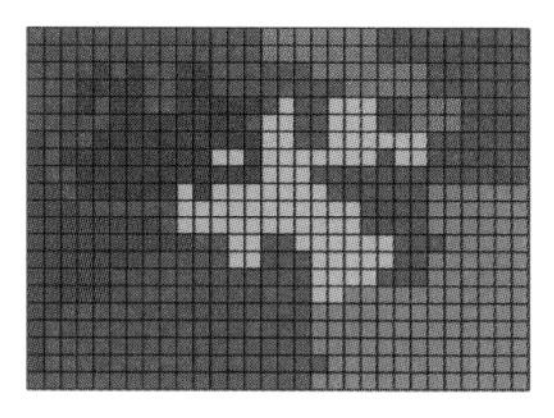

Raster dataset

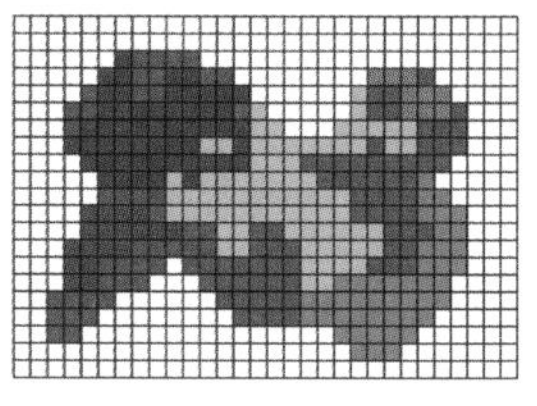

Analysis result using a mask

The cell size

The output cell size, or resolution, for any operation or function can be set to any size desired. The default output resolution is determined by the coarsest of the input raster datasets.

The cell size and analysis

Cells in different raster datasets do not need to be stored in the same resolution. But when processing between multiple datasets, the cell resolution, as is the case with the registration, needs to be the same. When multiple raster datasets are input into any Spatial Analyst function and their resolutions are different, one or more of the input datasets will be automatically resampled, using the nearest neighbor assignment (for additional information, see Chapter 4, 'Understanding raster data') to the coarsest input.

The nearest neighbor assignment resampling technique is used since it is applicable to both discrete and continuous value types, while bilinear and cubic are only applicable to continuous data. A resampling technique is necessary because rarely do the centers of the input cells align with the transformed cell centers of the desired resolution.

The default resampling to the coarsest resolution of the input rasters can be changed in the Cell Size tab of the Options dialog box to a specific cell size or to the minimum of the input raster datasets. Caution must be taken when specifying a finer cell size than the coarsest input because the resolution of the output cannot be more accurate than the coarsest of the inputs. Specifying a cell size of 50 meters when the input raster datasets are 100 meters creates an output raster with a cell size of 50 meters, but the accuracy is still 100.

When performing analysis, make sure you are asking appropriate questions of the cell size. That is, you will not study mouse movement when the cell size is five kilometers, and you will not want to use five-kilometer cells when studying the effects of global warming over the earth.

Handling projections during analysis

Raster datasets must be registered with one another before completing any analysis or processing between them. Each location on the ground must be represented by the same x,y cell address on the different input datasets. This means that the input raster datasets have to be in the same coordinate space or coordinate system—in the same projection. The coordinate space of the output will be dependent on the coordinate space of the input datasets. If two or more input raster datasets in an expression are not in the same coordinate space, Spatial Analyst will automatically put them in the same space using the following rules.

The default behavior:

If only one raster dataset is input, the output will be in the same coordinate space as the input—a very common situation.

If multiple raster and feature datasets are in the same coordinate space, the output will also be in that same coordinate space.

If more than one raster dataset is input, the output will be in the same coordinate space as the first input.

If feature and raster data with different coordinate spaces are input to the same function, the feature dataset will be projected to the coordinate space of the raster; the output will be in the coordinate space of the raster.

If feature data is input, the output will be in the same coordinate space as the first input.

Overriding the default:

On the General tab of the Options dialog box, you can set the coordinate space of all output raster datasets to be the same as that specified for the data frame.

Automatically transforming a raster or feature dataset into the common coordinate system in the cases identified above is referred to as projecting on the fly. To maintain the speed of on-the-fly projection, a low-order polynomial transformation is applied to the dataset. The on-the-fly projection transformation is less accurate than if you project the dataset using the geometric transformation tools that are available in ArcMap and Spatial Analyst, which are discussed in Chapter 4, 'Understanding raster data'.

Performing analysis

Section 3

Setting up your analysis environment 6

IN THIS CHAPTER

- **Creating temporary or permanent results**
- **Specifying a location on disk for the results**
- **Using an analysis mask**
- **Setting the coordinate system for results**
- **Setting the extent for results**
- **Specifying the cell size for results**

Specifying a certain extent, cell size, and working directory for your analysis results is a prerequisite to performing analysis. For instance, you may only be interested in analyzing a small piece of a geographic area, or you may want to write the results to a specific location.

Setting the options for your analysis results enables you to control the output directory for your results, the analysis extent, and the cell size. It also enables you to specify an analysis mask and a snapping extent, if appropriate. It is recommended that you set up your analysis options before you perform analysis on your data, or you can accept the defaults. By default, the directory for your analysis results is set to that of your system's temporary directory, usually c:\temp, the cell size is set to that of the largest cell size of your inputs, and the extent is set to the intersection of your input data.

This chapter will explain the following:

- Understanding and creating temporary and permanent results
- How to specify a location on disk for your analysis results
- What an analysis mask is and how to apply one
- How to set the extent for your analysis results
- The importance of cell size and how to specify this for your analysis results

Creating temporary or permanent results

By default, most results from analysis are temporary. Exceptions are the conversion functions and functions that do not output raster data. In these cases, results will be permanent by default.

Results from all other functions can be made permanent in three ways:

- By supplying a name for the result in a function dialog box
- By creating a temporary result, then making the temporary result permanent
- By saving the map document, which makes all temporary results permanent in the specified working directory, using the default output names

Tip

Setting your working directory

Click Options on the Spatial Analyst toolbar, then click the General tab to set up your working directory for your analysis results.

Making your results permanent within a function dialog box

1. When performing any function, type the name for the output and it will be permanently saved to your working directory.

   Alternatively, type a location on disk and a name for the output or use the Browse button to navigate to a folder on disk.

   Results will be permanent.

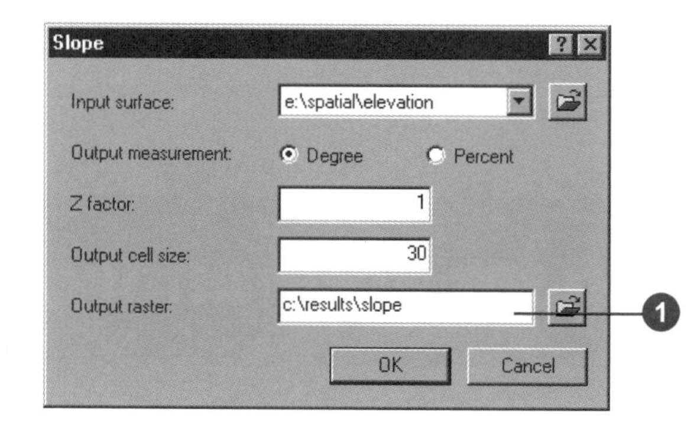

Making temporary results permanent

1. Right-click the temporary result in the table of contents and click Make Permanent.
2. Navigate to the directory in which you wish to save the result and specify a name.
3. Click Save.

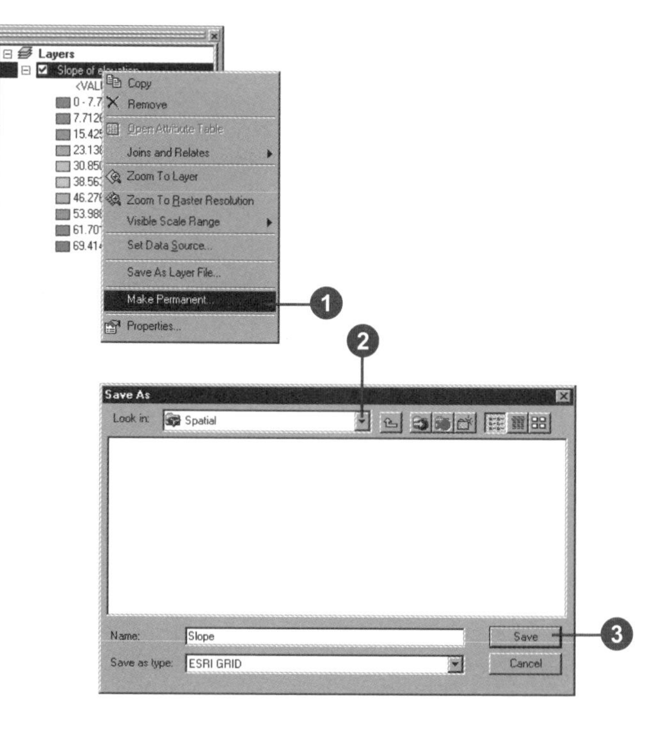

Tip

Why save your map document?

Saving the map document is a quick way to make all your temporary analysis results permanent and provides a way to save your work in order to continue your analysis at a later date.

Tip

A quick way to save your map document

Once you have specified a location and name for your map document, simply click the Save button on the Standard toolbar to save your work.

Making results permanent by saving the map document

1. Click the File menu and click Save As.
2. Navigate to the location in which you want to save the map document.
3. Type a filename.
4. Click the Save as type dropdown arrow and click ArcMap Documents (*.mxd).
5. Click Save.

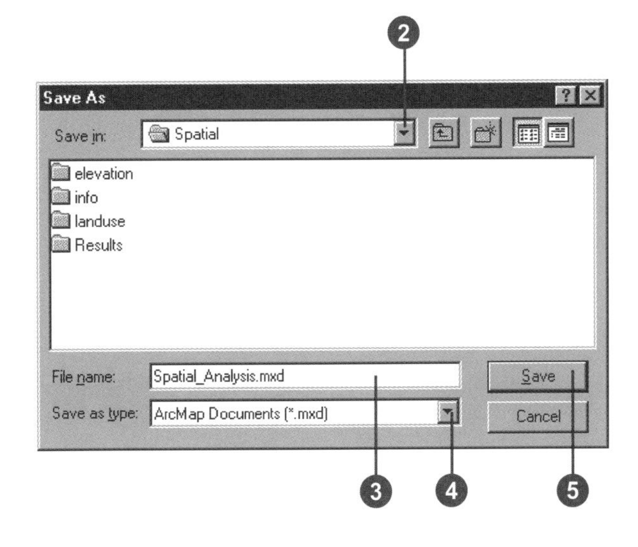

Specifying a location on disk for the results

The default location for your analysis results is your system's temporary directory, usually c:\temp.

There are two ways to specify where your analysis results should go. The first way is to specify a location on disk for your results using the Analysis Options dialog box before you perform any analysis. This way, all your analysis results will go to this directory. The second way is to specify a location on disk each time you perform analysis in each of the function dialog boxes. This is useful if you want to sort your analysis results into different folders.

Specifying a location for all analysis results via the Analysis Options

1. Click the Spatial Analyst dropdown arrow and click Options.
2. Click the General tab.
3. Type a location on disk for your analysis results or use the Browse button to navigate to a directory.
4. Click OK.

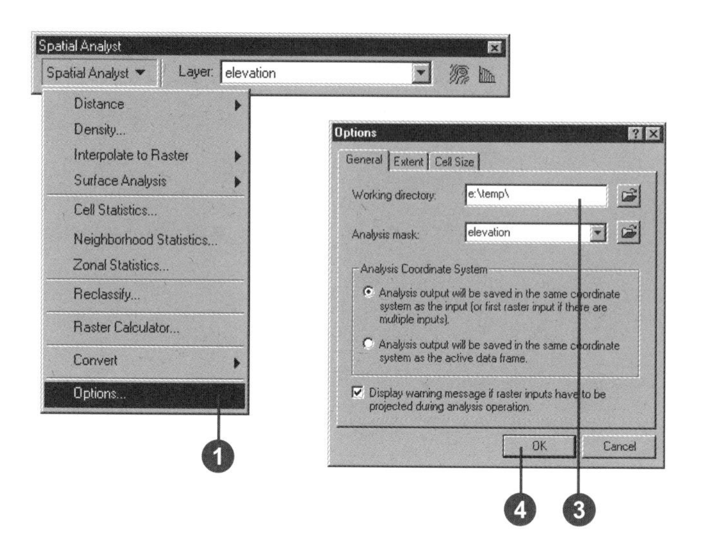

Specifying a location on disk for each output from a function

1. Type a location on disk and a name for the output when performing any function.

   Alternatively, use the Browse button to navigate to a folder on disk.

   By specifying a location and a name, the results will be permanent.

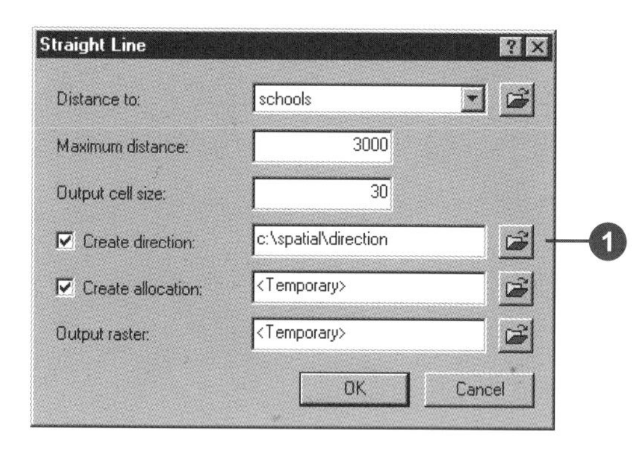

Tip

Creating a new working directory

Simply type the path to the new location in the Working directory input box.

Tip

Using your working directory

If you have set up your working directory, type the name for the output in a function dialog box to save it permanently to your working directory.

Using an analysis mask

Sometimes you only want to perform analysis on a selected set of cells and you want to mask out the rest.

Setting an analysis mask is a two-step process:

First, the analysis mask must be created, if you do not already have one. An *analysis mask* identifies those cells that will be considered when performing an operation or function. All NoData cells in the analysis mask will be masked out and assigned the NoData value on all subsequent output raster datasets. An analysis mask can be created via the Reclassify dialog box. Second, the analysis mask must be specified in the General tab of the Options dialog box in order for it to be used in all subsequent analyses.

Creating the analysis mask by reclassifying

1. Click the Spatial Analyst dropdown arrow and click Reclassify.
2. Click the Input raster dropdown arrow and click the raster from which you wish to create the analysis mask.
3. Click the Reclass field dropdown arrow and click the field you want to use.
4. Click the values you wish to exclude from any further processing.
5. Click Delete Entries.
6. Check the Change missing values to NoData check box.

   The values you deleted will be turned to NoData in the output raster.
7. Type a location on disk and a name for the mask.

   Alternatively, click the Browse button to browse to a folder in which to place the result.
8. Click OK.

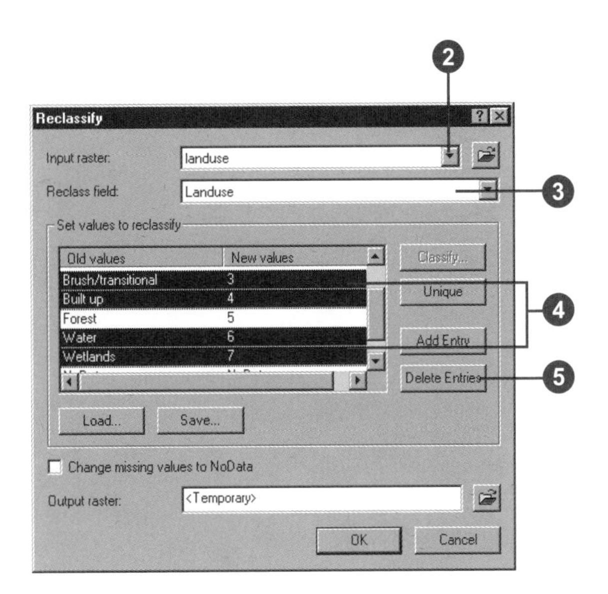

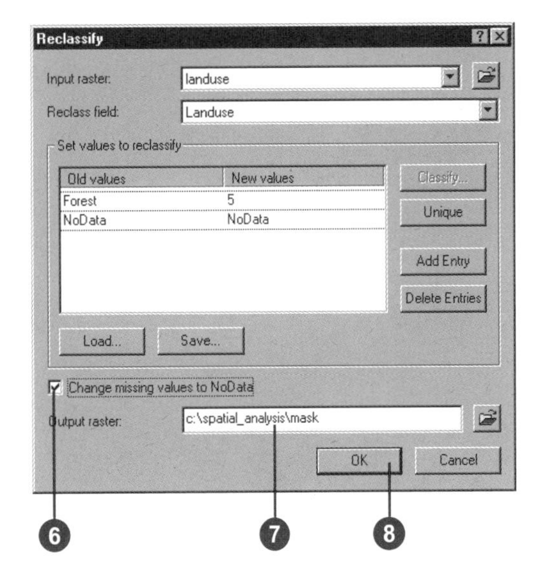

Tip

An alternative way to create an analysis mask

Rather than creating the analysis mask based on attributes, you can specify a spatial boundary as the mask. Create a new feature dataset in ArcCatalog, digitize the spatial boundary in ArcMap, then convert these features to a raster to create the mask. All NoData cells, areas outside of the original features, in the analysis mask will be set to the NoData value on all subsequent output raster datasets.

Tip

An alternative to setting an analysis mask

For layers for which you can access the table, right-click the layer in the table of contents and click Open Attribute Table. Select rows from the table. This selection will be respected when you perform a spatial function, so analysis will only be performed on the selected set.

Using the mask dataset in all subsequent analysis

1. Click the Spatial Analyst dropdown arrow and click Options.
2. Click the General tab.
3. Click the Analysis mask dropdown arrow and click the created mask.
4. Click OK.

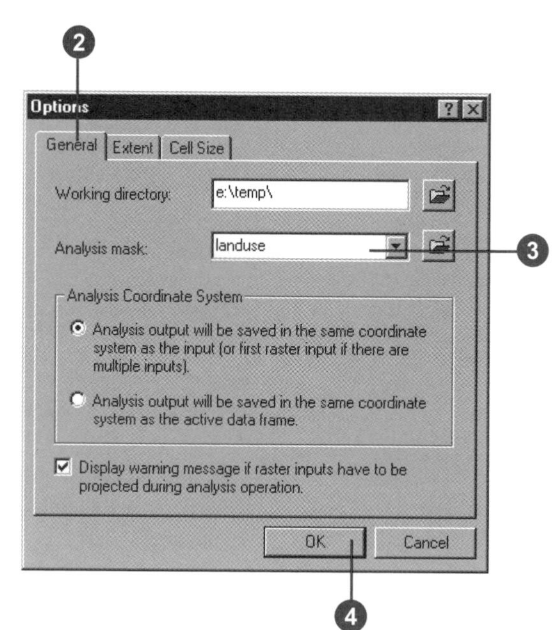

About the coordinate system and analysis

When performing analysis you can control the changing of coordinate systems.

To make on-the-fly projection of raster data for display fast, an approximation of the actual projection transformation is used and only the pixels that need to be drawn on the screen are transformed.

This transformation approximation can introduce error. It is well suited for use in low- to mid-latitudes and study areas of local or regional scale. It is not well suited for global or continental scale analysis. At high latitudes, it is only appropriate for small study areas, approximately two degrees of latitude and longitude at 60 degrees, and less than one degree at 75 degrees and higher. When working at the equator with cylindrical projections, areas as large as 10 degrees can be used with minimal error introduced by the projection.

Improved projection results can be achieved by using a rigorous cell-by-cell or piecewise reprojection of your data. Use the ARC commands PROJECT or PROJECTGRID, respectively.

Specifying the coordinate system option for analysis results

1. Click the Spatial Analyst dropdown arrow and click Options.
2. Click the General tab.
3. Click the Analysis Coordinate System option you wish to use.

   By default, analysis results will be saved in the coordinate system of the first input raster with a defined coordinate system. This minimizes the reprojection of raster data, which can be slow and introduce error.

   Alternatively, click the second option to have analysis results saved in the coordinate system of the data frame.
4. Click OK.

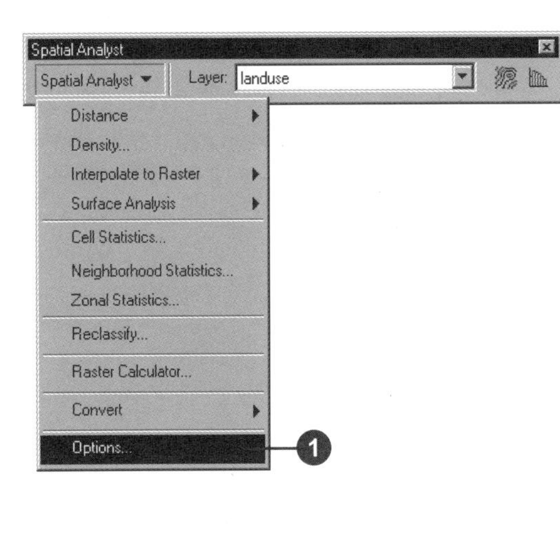

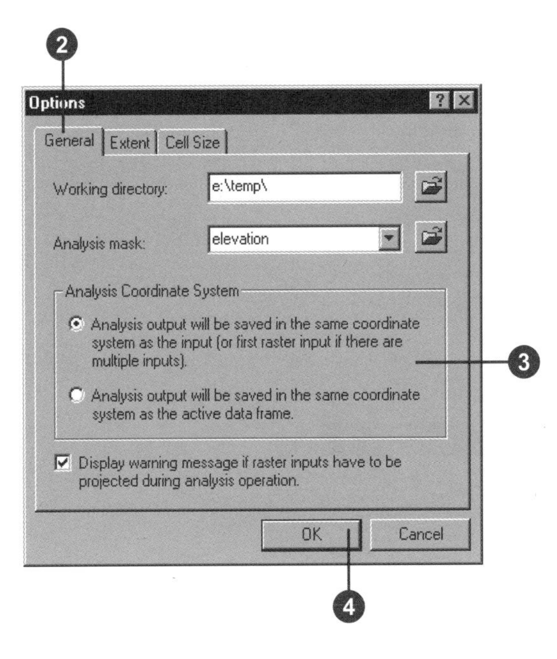

Setting the extent for results

The extent of a layer is the x,y coordinates for the bottom-left and the top-right corners. The analysis extent for your results can be set in the Extent tab of the Options dialog box. The default extent is set to Intersection of Inputs, so any analysis will only be performed where all layers overlay—the minimum of the inputs. You may wish to change the default setting.

Union of Inputs sets the extent of the results to be the same as the combined extent of inputs to a function. You may only wish to perform analysis on the area visible in the map—Same as Display—or you may wish to make the extent the same as another layer in the table of contents—Same as Layer "filename". Alternatively, you can specify a custom extent—As Specified Below.

Tip

Setting the snap extent

Setting the snap extent to a specific raster dataset will snap all output raster datasets to the cell registration of the specified raster dataset.

Specifying an extent for analysis results

1. Click the Spatial Analyst dropdown arrow and click Options.
2. Click the Extent tab.
3. Click the Analysis extent dropdown arrow and choose an option to specify the extent for all subsequent analysis results.
4. Click OK.

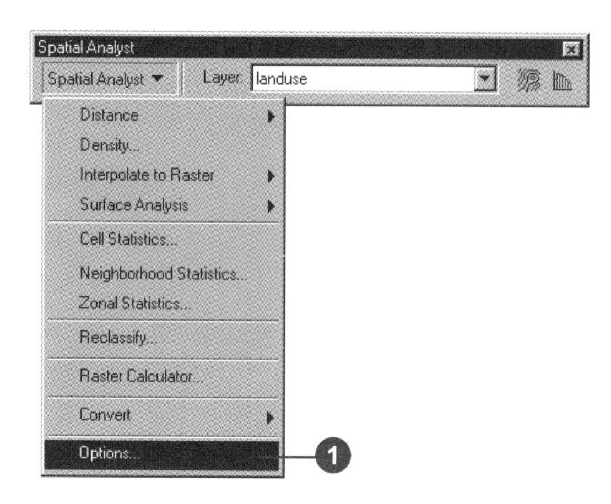

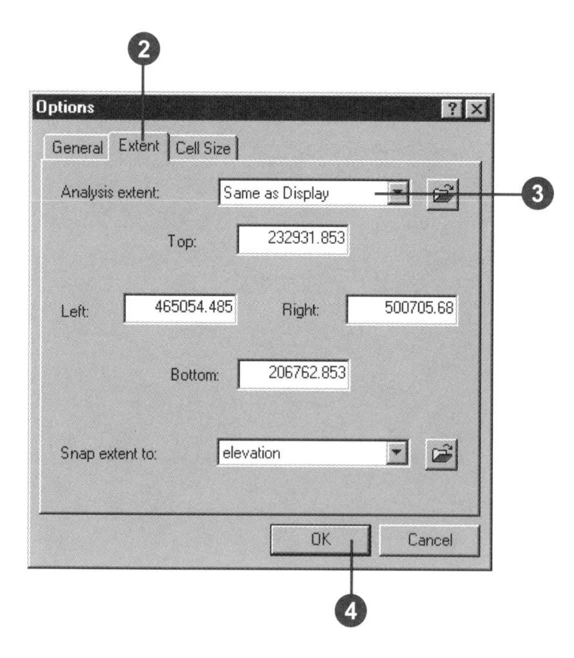

Setting the cell size for results

The default cell size, or resolution, for analysis results is set to the input raster dataset with the largest cell size—the Maximum of Inputs.

The default cell size when a feature dataset is used as input to a function is to take the width or the height—whichever is shortest—of the extent of the input feature dataset and divide by 250 to get 250 cells.

Exercise caution when specifying a cell size finer than the input raster datasets. No new data is created; cells are interpolated using nearest-neighbor resampling. The result is as precise as the coarsest input.

The default cell size can be changed on the Cell Size tab of the Options dialog box. The cell size you specify will be applied to all subsequent results.

Other options available: Minimum of Inputs sets the cell size of your analysis results to the input raster dataset with the smallest cell size, As Specified Below enables you to specify a cell size for analysis results, and Same As Layer enables you to select an input raster layer on which to base the cell size of your analysis results. ▸

Specifying a cell size for all subsequent analysis results

1. Click the Spatial Analyst dropdown arrow and click Options.
2. Click the Cell Size tab.
3. Click the Analysis cell size dropdown arrow and choose the appropriate option.
4. Click OK.

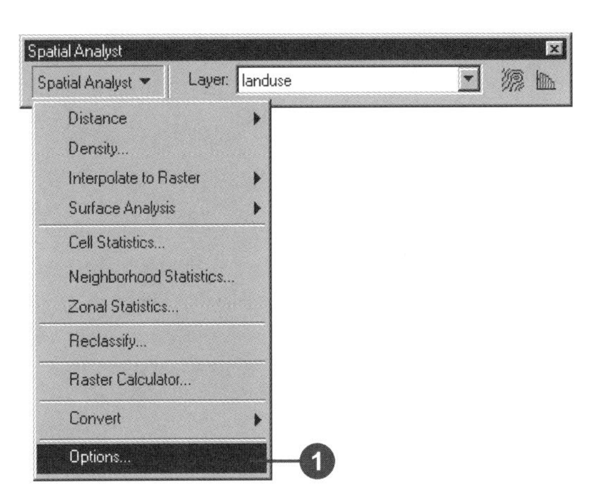

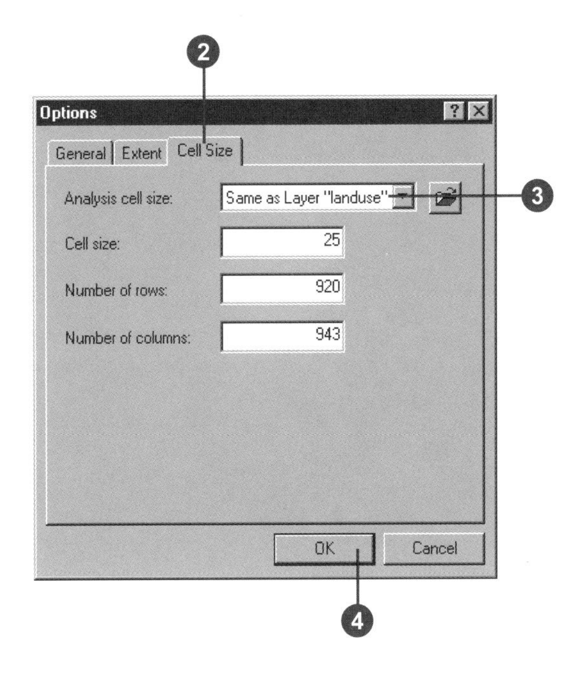

Alternatively, specify the number of rows and columns to split your analysis extent into, and an appropriate cell size will be applied.

For functions that accept nonraster data, you can specify the cell size for your output raster dataset directly in the function dialog box. The default is whatever is set on the Cell Size tab of the Options dialog box—be this the default or a cell size you specified.

Tip

Finding out a raster layer's cell size

To find out the cell size of a raster layer, right-click the raster layer in the table of contents, click Properties, then click the Source tab.

Applying a different cell size than the default for certain functions

1. Type a cell size.
2. Click OK.

   The cell size you specified will be applied to your Output raster.

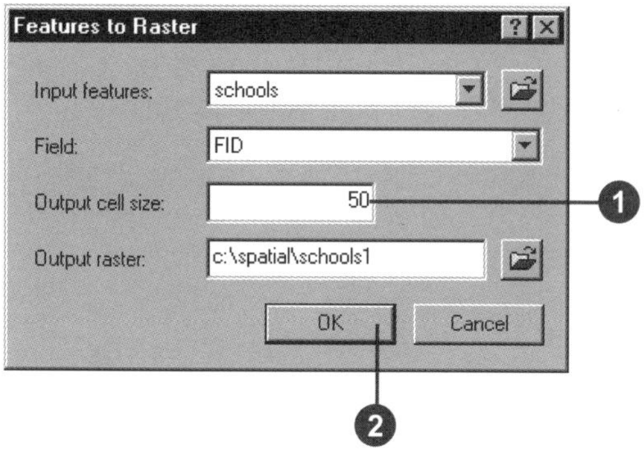

Features to Raster

Input features:	schools
Field:	FID
Output cell size:	50
Output raster:	c:\spatial\schools1

OK | Cancel

Performing spatial analysis

7

IN THIS CHAPTER

- **Mapping distance**
- **Mapping density**
- **Interpolating to raster**
- **Performing surface analysis**
- **Calculating cell statistics**
- **Calculating neighborhood statistics**
- **Calculating zonal statistics**
- **Reclassifying your data**
- **Using the raster calculator**
- **Converting your data**

Spatial Analyst provides you with tools to perform spatial analysis on your data that help you solve your spatial problems.

The previous chapter gave you information about setting the analysis properties before performing analysis. This chapter will provide you with detailed information about the analytical functions in Spatial Analyst, explaining what each of these functions do, why you might want to use them, and how to use these functions to perform tasks.

The Spatial Analyst functions accept layers added to ArcMap, and raster or feature datasets that you can browse to in each function dialog box. The Spatial Analyst functions also support selection on layers, so you can select certain values in an attribute table or on the map and use this selection in your analysis.

This chapter is organized in the order of the functions on the user interface, so if you want, for instance, information on converting your data—near the bottom of the pulldown menu—simply visit the last few pages of this chapter to get more information.

This chapter contains:

- Conceptual information about each function
- Step-by-step details of how to use each function

Use this chapter as a reference guide, looking up a particular function when you need more information.

Mapping distance

What are distance mapping functions?

The distance mapping functions are *global functions*. They compute an output raster dataset where the output value at each location is potentially a function of all the cells in the input raster datasets.

There are several distance mapping tools for measuring both straight line—Euclidean—distance and distance measured in terms of other factors, such as the cost to travel over the landscape. The outputs from the Straight Line Distance functions are normally used directly, while the outputs from the Cost Weighted Distance functions are most commonly used to compute shortest, or least-cost, paths.

Straight Line Distance functions

The *Straight Line Distance* function measures the straight line distance from each cell to the closest source—the source identifies the objects of interest, such as wells, roads, or a school. The distance is measured from cell center to cell center.

The *Straight Line Allocation* function assigns each cell the value of the source to which it is closest. The nearest source is determined by the Straight Line Distance.

The *Straight Line Direction* function computes the direction to the nearest source, measured in degrees.

Cost Weighted Distance functions

The *Cost Weighted Distance* function modifies the Straight Line Distance by some other factor, which is a cost to travel through any given cell. For example, it may be shorter to climb over the mountain to the destination, but it is faster to walk around it.

The *Cost Weighted Allocation* function identifies the nearest source cell based on accumulated travel cost.

The *Cost Weighted Direction* function provides a road map, identifying the route to take from any cell, along the least-cost path, back to the nearest source.

The Distance and Direction raster datasets are normally created to serve as inputs to the pathfinding function, the *shortest*, or least-cost, *path*.

Why is it useful to map distance?

By mapping distance, you can find out information, such as the distance to the nearest hospital from certain areas for an emergency helicopter, or find all fire hydrants within 500 meters of a burning building. Alternatively, you could find the shortest, or least-cost, path from one location to another, based on some cost factor.

The pages that follow explain Straight Line Distance, Allocation, Cost Weighted Distance, and Shortest Path in more detail.

Straight line distance

What are the Straight Line Distance functions?

The Straight Line Distance functions describe each cell's relationship to a source or a set of sources.

There are three potential outputs from this function.

Primary output:

- **Straight Line Distance** gives the distance from each cell in the raster to the closest source.

  Example of usage: What is the distance to the closest town?

Optional outputs:

- **Straight Line Allocation** identifies the cells that are to be allocated to a source based on closest proximity.

  Example of usage: Which town am I closest to?

- **Straight Line Direction** gives the direction from each cell to the closest source.

  Example of usage: What is the direction to the closest town?

The source

The source identifies the location of the objects of interest, such as wells, shopping malls, roads, forest stands, and so on. If the source is a raster, it must contain only the values of the source cells—all other cells must be NoData. If the source is a feature, it will internally be transformed into a grid when you run the function.

The straight line distance raster

The straight line distance raster contains the measured distance from every cell to the nearest source. The distances are measured in projection units, such as feet or meters, and are computed from cell center to cell center.

The Straight Line Distance function is used frequently as a standalone function for applications, such as finding the nearest hospital for an emergency helicopter. Alternatively, this function can be used when creating a suitability map, when you need to include data representing the distance from a certain object (for more details, see 'Finding a site for a new school in Stowe, Vermont, USA' in Chapter 2).

In the example below, the distance to each town is found. This sort of information could be extremely useful for planning a hiking trip. You may want to stay within a certain distance of a town in case of emergencies, or know how much further you have to travel to pick up supplies.

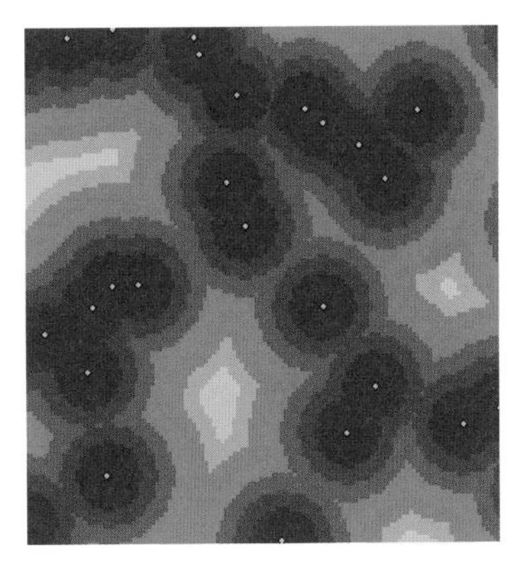

The straight line distance to the nearest town from every location.

Optional outputs

The Straight Line Allocation raster dataset

Every cell in the Straight Line Allocation raster is assigned the value of the source to which it is closest. The nearest source is determined by the Straight Line Distance. Use this function to assign space to objects, such as identifying the customers served by a group of stores.

In the example below, the allocation function has identified the town that is closest to each cell. This could be valuable information if you need to get to the nearest town from a remote location (for more information, see 'Allocation' in this chapter).

Allocating cells to sources: Which areas are served by which town?

The Straight Line Direction raster dataset

The Straight Line Direction raster contains the azimuth direction from each cell to the nearest source. The directions are measured in degrees, where north is 0 degrees.

In the example below, the direction to the nearest town is found from every location. This could provide useful information for an emergency helicopter when transporting an injured hiker to the nearest town for medical treatment.

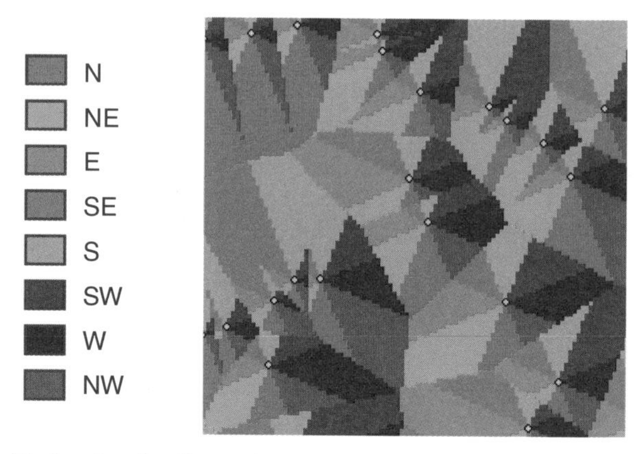

Finding the direction to the nearest source: What is the direction from this location to the nearest town?

Keep in mind that the Straight Line Distance functions give you information according to Euclidean, or straight line, distance.

It may not be possible to travel in a straight line to get to a location; you may have to avoid obstacles, such as a river or a steep slope. In such cases, you should consider using the Cost Weighted Distance function to achieve more realistic results (see 'Cost Weighted Distance' later in this chapter).

Straight line distance

The Straight Line Distance function enables you to calculate how far each cell is from the nearest source. The source can be anything you choose, from a well, to a road, to a group of retail stores, and can be in any supported raster or feature *format*.

Tip

Setting analysis options

Click Options on the Spatial Analyst toolbar to set up your working directory, extent, and cell size for your analysis results.

Tip

Browsing for files or directories

If the file you need is not in your table of contents, or if you need to check the directory to place your results, click the Browse button.

Tip

Deciding on the maximum distance

Use the Measure tool on the Tools toolbar to decide on the maximum distance from each source.

Calculating straight line distance

1. Click the Spatial Analyst dropdown arrow, point to Distance, and click Straight Line.
2. Click the Distance to dropdown arrow and click the layer to find the distance to.
3. Optionally, specify a maximum distance. Cells outside this distance will not be considered in the calculation and will be given the value of NoData.

   Leaving the Maximum distance blank will not put a limit on how far distances will be measured.
4. Specify an output cell size for the result(s).
5. Optionally, click Create direction to create a raster displaying the straight line direction to the closest source.
6. Optionally, click Create allocation to create a raster where every cell is assigned the value of the closest source.
7. Type a name for the result or leave the default to create a temporary result in your working directory.
8. Click OK.

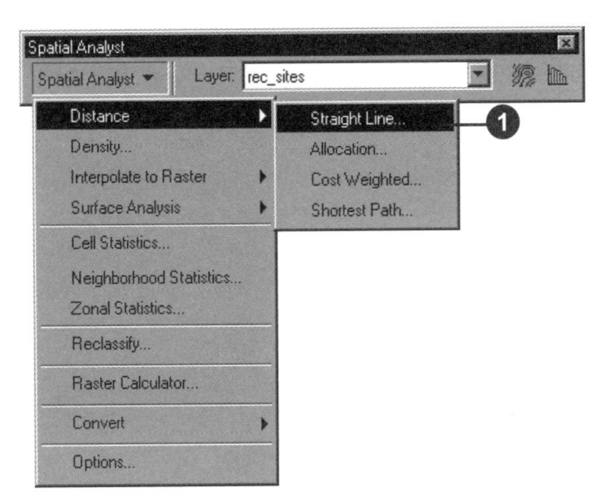

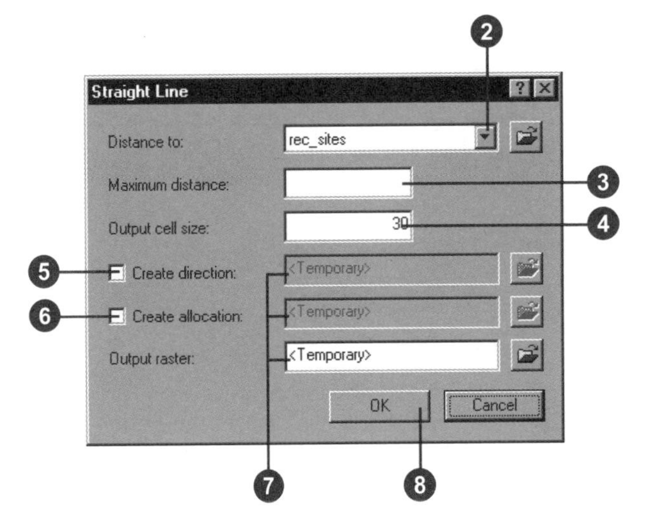

Allocation

What is the Allocation function?

The Allocation function allows you to identify which cells belong to which source based on closest proximity, in a straight line.

An output raster is produced that records the identity of the closest source cell for each cell. Each cell in an allocation raster receives the value of the source cell to which it will be allocated.

	1	1			
		1			
2					

1	1	1	1	1	1
1	1	1	1	1	1
1	1	1	1	1	1
2	1	1	1	1	1
2	2	2	1	1	1
2	2	2	2	2	1

= NoData

Note that the Allocation function can also be performed via the Straight Line Distance function or the Cost Weighted Distance function.

Performing the Allocation function via the Straight Line Distance function allows you to find the cells that are to be allocated to which source based on closest proximity, in a straight line.

Performing the Allocation function via the Cost Weighted Distance function takes the cost of traveling over the land into consideration rather than the straight line distance (see 'Cost weighted distance' later in this chapter).

Why use the Allocation function?

Use the Allocation function to perform analyses such as:

- Identifying the customers served by a series of stores
- Finding out which hospital is the closest
- Finding areas with a shortage of fire hydrants
- Locating areas that are not served by a chain of supermarkets

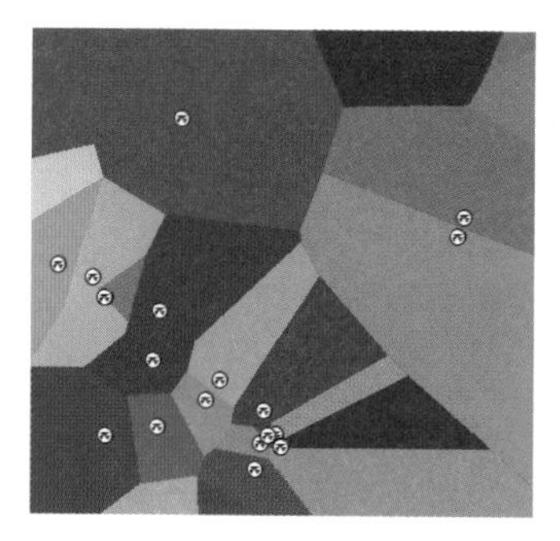

The example to the left identifies the areas of land supported by a recreation site. You can easily identify the areas that may be in need of more recreation sites—mainly areas in the northeast half of the raster.

Allocating cells to sources

The Allocation function allows you to allocate cells to the closest source. The source can be anything you choose, such as a point dataset displaying the location of parks, and can be in any supported raster or feature format.

Tip

Deciding on the Maximum Distance

Use the Measure tool on the Tools toolbar to decide what the maximum distance from each source should be.

Tip

Setting analysis options

Click Options on the Spatial Analyst toolbar to set up your working directory, extent, and cell size for your analysis results.

Tip

Browsing for files or directories

If the file you need is not in your table of contents, or if you need to check the directory to place your results, click the Browse button.

Calculating straight line allocation

1. Click the Spatial Analyst dropdown arrow, point to Distance, and click Allocation.
2. Click the Assign to dropdown arrow and click the layer containing sources to which you wish to assign cells.
3. Optionally, specify a maximum distance. Cells outside this distance will not be considered in the calculation and will be given the value of NoData.

   Leaving the Maximum distance blank will not put a limit on how far distances will be measured.
4. Specify an Output cell size for the result—the default cell size is that specified in the Options dialog box.
5. Type a name for the result or leave the default to create a temporary result.
6. Click OK.

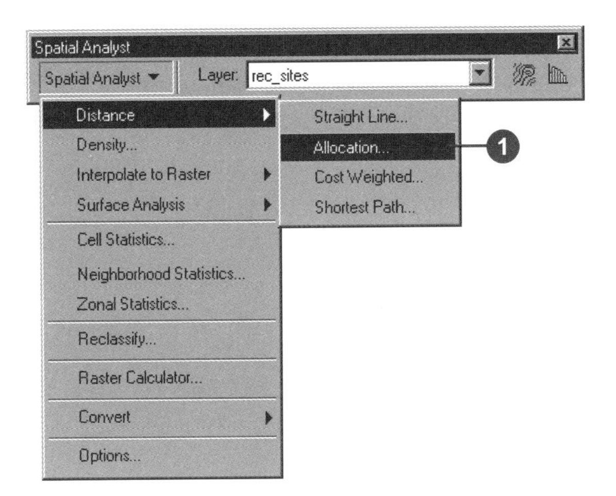

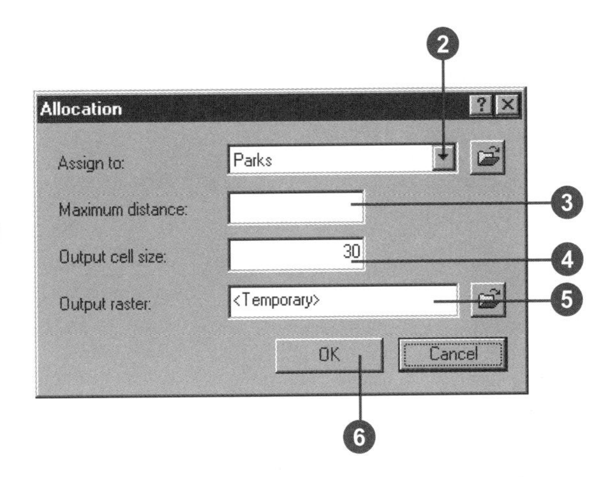

Cost weighted distance

What is cost weighted distance mapping?

Cost weighted distance mapping finds the least accumulative cost from each cell to the nearest, cheapest source. Cost can be money, time, or preference.

The functions that perform cost weighted distance mapping are similar to the Straight Line Distance functions, but instead of calculating the actual distance from one point to another, they compute the accumulative cost of traveling from each cell to the nearest source, based on the cell's distance from each source and the cost to travel through it—for example, it is easier to walk through a meadow than a swamp.

Why use the Cost Weighted Distance function?

Cost weighted distance modeling is useful whenever movement is based on geographic factors, such as animal migration studies or consumer travel behavior. Cost weighted distance may also be used to minimize construction costs for routing new roads, transmission lines, or pipelines.

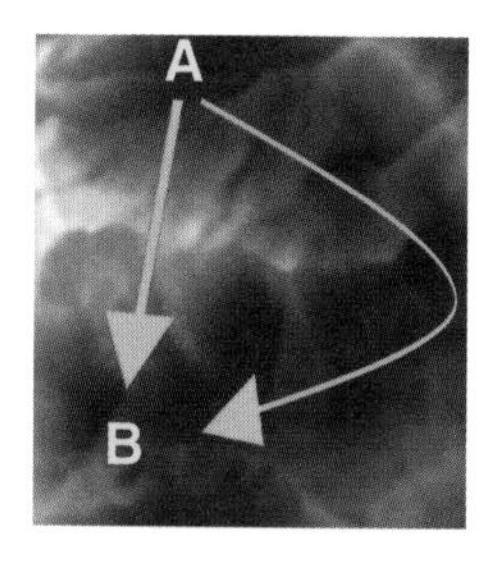

The straight line distance between two points is not necessarily the best. In the graphic to the left, the shortest path over the mountain takes three hours. The longer path around only takes two hours. If time were a cost, then the route with the longer distance should be taken. However, the aim may be to climb over the mountain. Applying cost weighted distance enables you to specify preferences in your input data. It may, for example, take longer to travel over the mountain due to steep slopes, so steep slopes should be given a higher cost when finding a suitable path from A to B.

Example: Finding the least-cost route for a road

In the following example, the Cost Weighted Distance functions are used to find the least-cost path for a new road. The Cost Weighted Distance function is the prerequisite to the Shortest Path function, which is discussed in the next section. The Shortest Path function determines the actual route for the road.

To calculate the least accumulative cost from each cell to the nearest source, the Cost Weighted Distance function needs a source and a cost raster.

The source

The source, as you can see in the graphic below, in red, is the starting point for the proposed road.

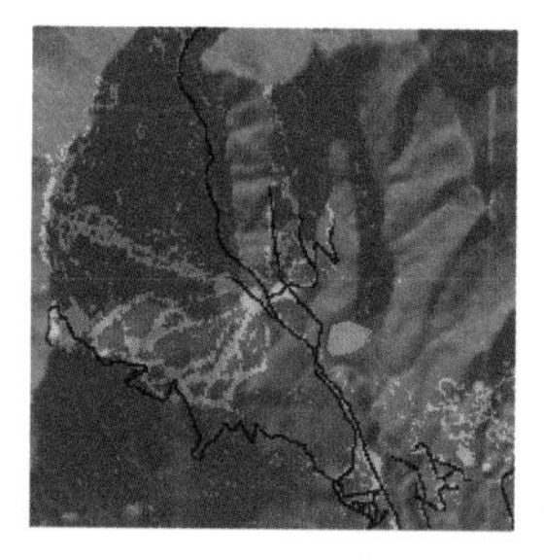

The cost raster

The cost raster identifies the cost of traveling through every cell. To create this raster, you need to identify the cost of constructing a road through each cell. Although the cost raster is a single dataset, it is often used to represent several criteria. In the following example, landuse and slope influence the construction costs. These datasets are in different measurement systems—landuse type and percent slope—so they cannot be compared relative to one another and must be reclassified to a common scale.

Creating a cost raster

Reclassifying your datasets to a common scale

In this example, slope and landuse have been reclassified on a scale of 1–10. The attributes of each dataset should be examined, in turn, to decide on their contribution to the cost of building a road. For example, it is more costly to traverse steep slopes, so steeper slopes will be assigned higher costs when reclassifying this dataset. The graphics below display the results.

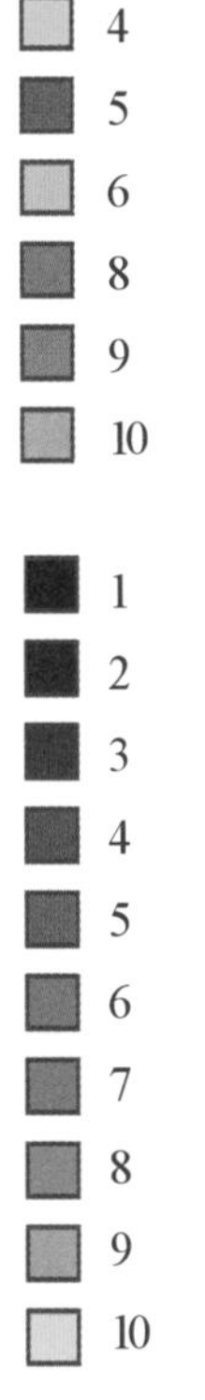

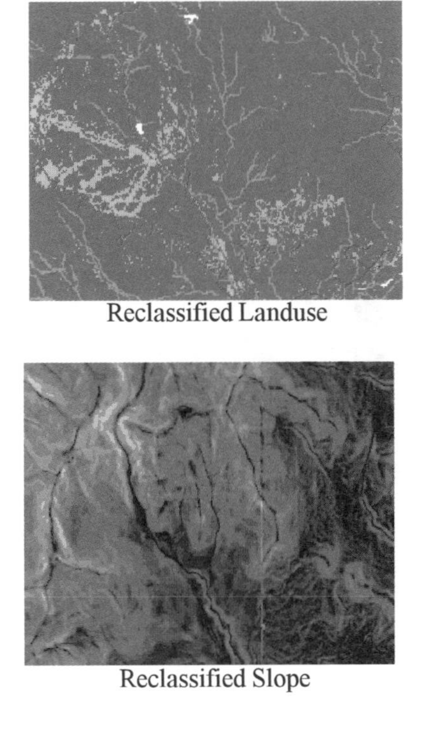

Reclassified Landuse

Reclassified Slope

High cell values are the more costly cells through which to route the road.

Weighting datasets according to percent influence

The next step in producing the cost raster is to add the reclassified datasets together. The simplest approach is to just add them together. However, you may know that some factors are more important than others. For instance, avoiding steep slopes may be twice as important as the landuse type, so you might, for example, give this dataset an influence of 66 percent and the landuse dataset an influence of 34 percent, to make 100 percent. The following diagram shows the conceptual process:

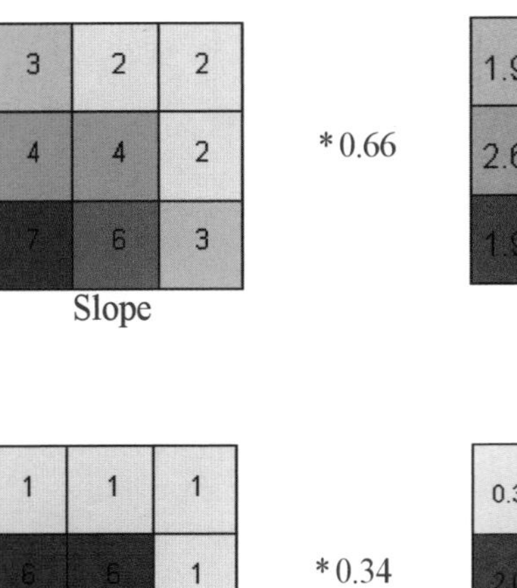

Combining the datasets

The final cost raster is the result of adding the weighted datasets together.

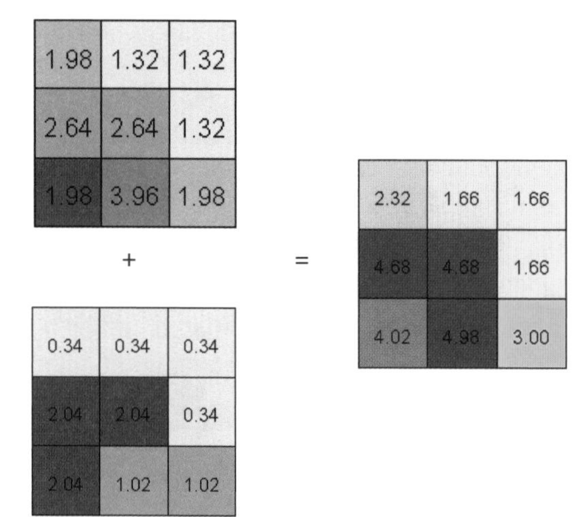

Taking this example, the following diagram shows the final cost raster, the result of reclassifying the datasets of slope and landuse, weighting each by 0.66 and 0.34, respectively, then combining the weighted datasets.

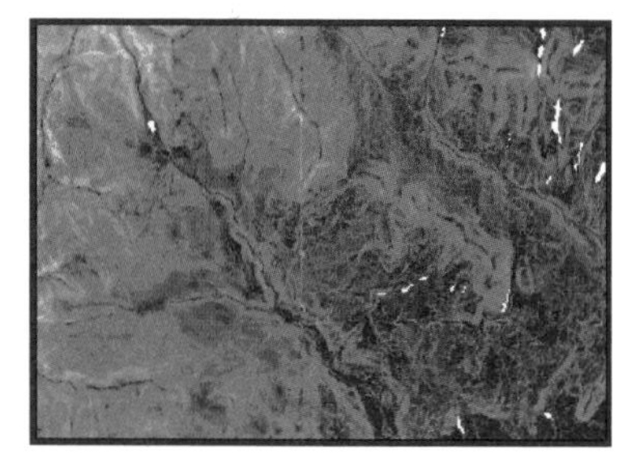

The cells shaded dark blue are the most suitable cells through which to route the road, as they are the least costly.

The Cost Weighted Distance function

Using the cost raster and the source, the Cost Weighted Distance function produces an output raster in which each cell is assigned a value that is the least accumulative cost of getting back to the source.

Using our example, the function takes the cost raster and calculates a value for each cell in the output cost-weighted distance raster that is the accumulated least cost of getting from that cell to the nearest source.

Every cell in the cost-weighted distance raster is assigned a value that represents the sum of the minimum travel costs that would be incurred by traveling back along the least-cost path to its nearest source.

In the example below, the accumulated least costly way of getting from the cell colored dark red to the school is 10.5.

5.0	7.5	10.5
2.5	5.7	6.4
	1.5	3.5

Two additional outputs—direction and allocation rasters—can be created from the Cost Weighted Distance function. These are explained on the following pages.

Direction

The cost-weighted distance raster tells you the least accumulated cost of getting from each cell to the nearest source, but it doesn't tell you which way to go to get there. The direction raster provides a road map, identifying the route to take from any cell, along the least-cost path, back to the nearest source.

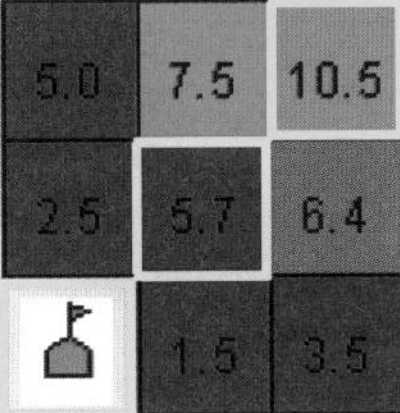

Cost Weighted

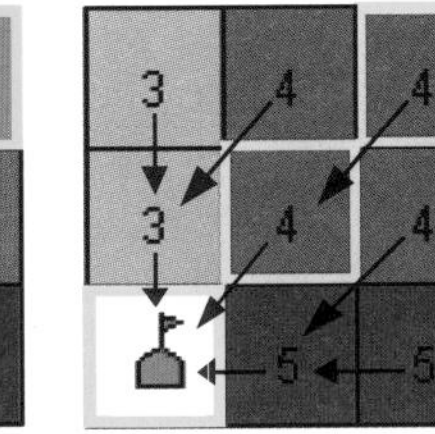

Direction

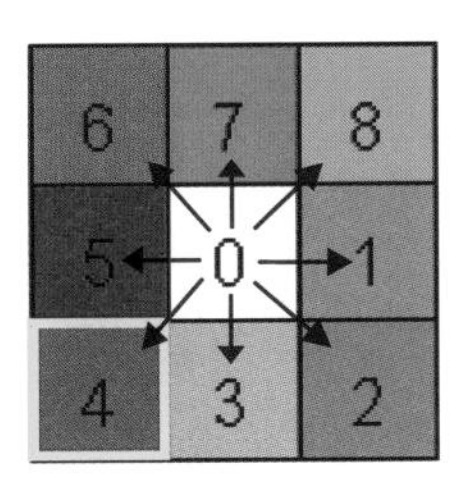

Direction Coding

The algorithm for computing the direction raster assigns a code to each cell that identifies which one of its neighboring cells is on the least-cost path back to the nearest source. In the direction coding diagram above, 0 represents every cell in the cost-weighted distance raster. Each cell is assigned a value representing the direction of the nearest, cheapest cell on the route of the least costly path to the nearest source.

For example, in the graphic above, the cheapest way to get from the cell with a value of 10.5 is to go diagonally, through the cell with a value of 5.7, to the source, the school site. The direction algorithm assigns a value of 4 to the cell with a value of 10.5, and 4 to the cell with a value of 5.7, because this is the direction of the least-cost path back to the source from each of these cells. This process is done for all cells in the cost-weighted distance raster to produce the direction raster, which tells you the direction to travel from every cell in the cost-weighted distance raster back to the source.

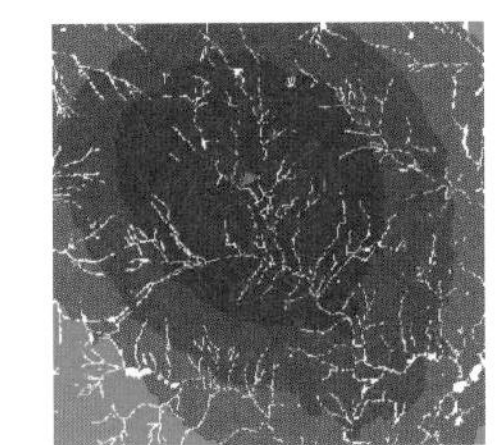

Cost Weighted Distance

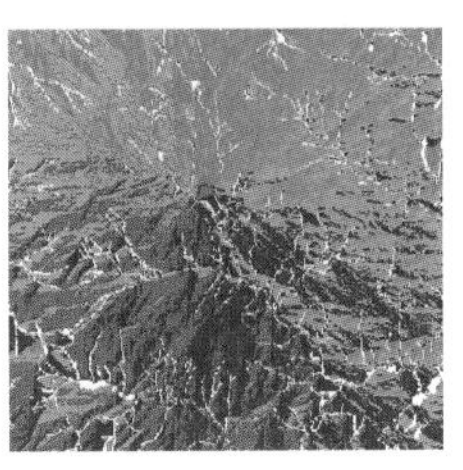

Direction

Both the cost-weighted distance and direction rasters are required if you want to go on to calculate the least-cost—shortest—path between source locations and destination locations.

Allocation

The cost-allocation raster identifies the nearest source from each cell in the cost-weighted distance raster. It is conceptually similar to the Straight Line Distance Allocation function, where each cell is assigned to its nearest source cell. However, near is expressed in terms of accumulated travel cost.

1	1	1
1	1	1
	1	1

All cells are allocated to the school source.

Cost weighted distance

The Cost Weighted Distance function calculates a value for every cell that is the least accumulated cost of traveling from each cell to the source.

The source can be anything you choose, such as a point layer displaying the location of a road junction or a building, and can be in any supported raster or vector format.

The Maximum Distance option allows you to specify that the processing is only done on cells up to a certain distance away from each source.

Tip

Setting analysis options

Click Options on the Spatial Analyst toolbar to set up your working directory, extent, and cell size for your analysis results.

Tip

Browsing for files or directories

If the file you need is not in your table of contents, or if you need to check the directory to place your results, click the Browse button.

Calculating cost weighted distance

1. Click the Spatial Analyst dropdown arrow, point to Distance, and click Cost Weighted.
2. Click the Distance to dropdown arrow and click the source layer.
3. Click the Cost raster dropdown arrow and click the raster to use.
4. Optionally, specify a Maximum distance. Cells outside this distance will not be considered in the calculation and will be given the value of NoData.

   Leaving the Maximum distance blank will not put a limit on how far distances will be measured.
5. Specify an Output cell size for the result.
6. Optionally, check Create direction to create a direction raster. This is a required input for the Shortest Path function.
7. Optionally, check Create allocation to create an Allocation raster.
8. Type a name for each result or leave the default to create a temporary result.
9. Click OK.

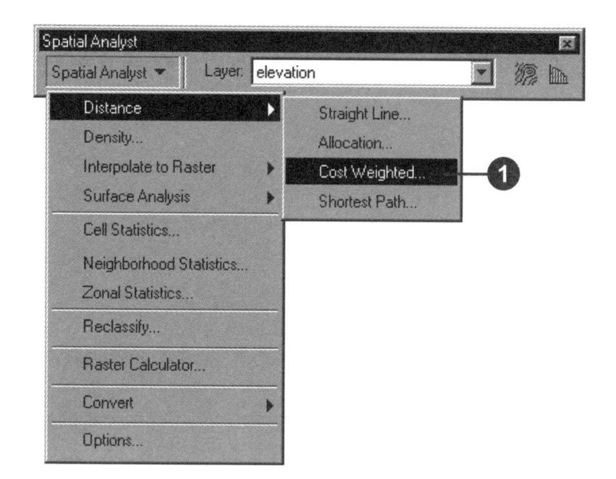

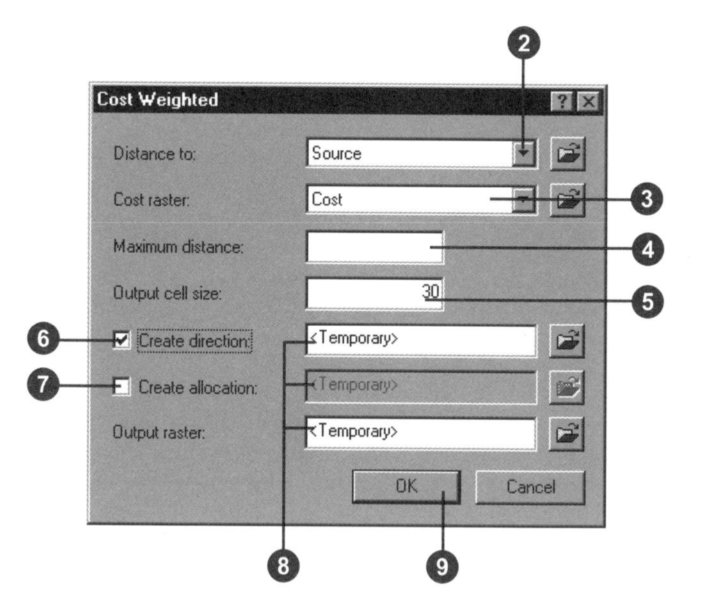

Shortest path

What is the Shortest Path function?

The Shortest Path function determines the path from a destination point to a source. Once you have performed the Cost Weighted Distance function, creating distance and direction rasters, you can then compute the least-cost—or shortest—path from a chosen destination to your source point, which in our original example was the starting point for the new road.

Why find the shortest path?

The shortest path travels from the destination to the source and is guaranteed to be the cheapest route—relative to the cost units defined by the original cost raster. Use it to find the best route for a new road in terms of construction costs, or to identify the path to take from several suburban locations—sources—to the closest shopping mall—destinations.

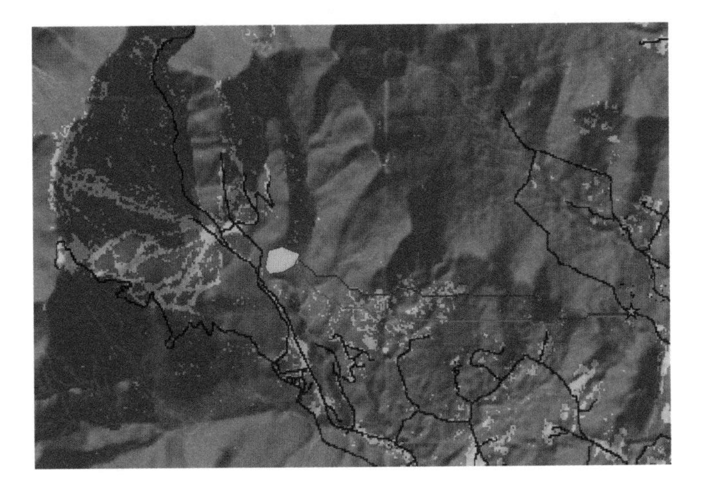

You can see two potential paths for the new road in the diagram above, in purple and red, to illustrate an important point. The purple line represents the path created using a cost raster where each input raster—landuse and slope—had the same influence. The red line represents the path created using a cost raster where the slope input raster had a weight, or influence, of 66 percent. By giving the slope input raster a higher weight, more attention was given to avoiding steeper slopes in the red path.

The point to understand from this is that it is important to spend time considering how to weight the rasters that make up the cost raster. How you weight your rasters depends on your application and the results you wish to achieve.

Finding the shortest path

The Shortest Path function finds the shortest, or least-cost, path from a source, or a set of sources, to a destination, or set of destinations, such as finding the least-cost path from several suburban locations—sources—to the closest shopping mall—destinations.

The Path type indicates the number of paths that will be found:

A path For Each Cell finds a path for each cell in each zone—each cell in every suburb receives its own path.

A path For Each Zone finds the one, least-cost path for each zone—each suburb receives only one path.

The Best Single path finds the least-cost path for all the zones—only the shortest path between one suburb and one mall is computed.

See Also

See 'Cost weighted distance' earlier in this chapter.

Performing shortest path

1. Click the Spatial Analyst dropdown arrow, point to Distance, and click Shortest Path.
2. Click the Path to dropdown arrow and click your destination layer.
3. Click the Cost distance raster dropdown arrow and click the raster you want to use.
4. Click the Cost direction raster dropdown arrow and click the raster you want to use.
5. Click the Path type dropdown arrow and choose an option, depending on how many paths you wish to be found.
6. Type a name for the result or leave the default to create a temporary result.
7. Click OK.

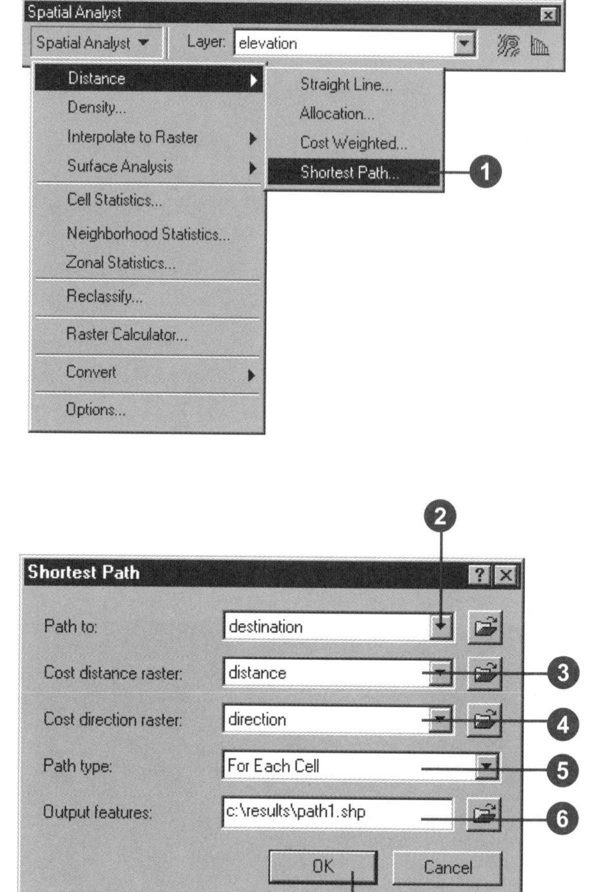

Mapping density

What is density?

By calculating *density* you spread point values over a surface. The magnitude at each sample location, line or point, is distributed throughout a landscape, and a density value is calculated for each cell in the output raster.

Density maps are predominantly created from point data, and a circular search area is applied to each cell in the output raster being created. The search area determines the distance to search for points in order to calculate a density value for each cell in the output raster.

Density calculations

You can calculate density using simple or kernel calculations.

In a simple density calculation, points or lines that fall within the search area are summed and then divided by the search area size to get each cell's density value.

The kernel density calculation works the same as the simple density calculation, except the points or lines lying near the center of a raster cell's search area are weighted more heavily than those lying near the edge. The result is a smoother distribution of values.

Why map density?

Density surfaces are good for showing where point or line features are concentrated. For example, you might have a point value for each town, representing total population, but you want to learn more about the spread of population over the region. With census data, you may have a point representing the number of people in each town. Since all the people in each town do not live at the population point, by calculating density you can create a surface showing the predicted distribution of the population throughout the landscape.

The following graphic gives an example of a density surface. When added together, the population values of all the cells equal the sum of the population of the original point layer.

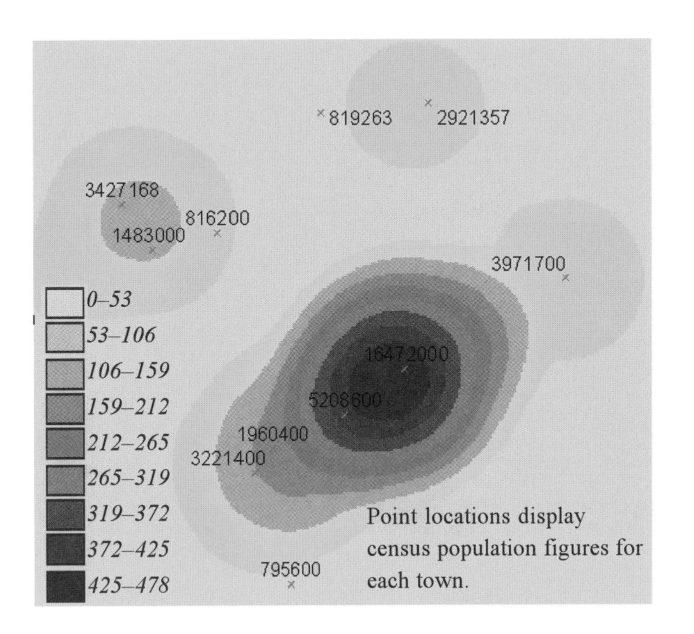

Density

The Density function allows you to create a continuous density surface from a set of input features. It can provide a more realistic interpretation of your values—point values are spread out, giving you a better indication as to their distribution over a surface.

Tip

Deciding on the search radius

Click the Measure tool on the Tools toolbar and measure distance from a certain point. The distance is reported in the status bar. It will help you reach a decision on how big to make the search radius.

Tip

Setting analysis options

Click Options on the Spatial Analyst toolbar to set up your working directory, extent, and cell size for your analysis results.

Calculating density

1. Click the Spatial Analyst dropdown arrow and click Density.
2. Click the Input data dropdown arrow and click the input layer.
3. Click the Population field dropdown arrow and click the field you want to use.
4. Click either Kernel or Simple Density type.
5. Type a value in the Search radius text box to determine the distance to search for points or lines from each cell in the output raster.
6. Click the Area units dropdown arrow and choose the units in which the density values should be presented.
7. Specify an Output cell size.
8. Type a name for the result or leave the default to create a temporary result.
9. Click OK.

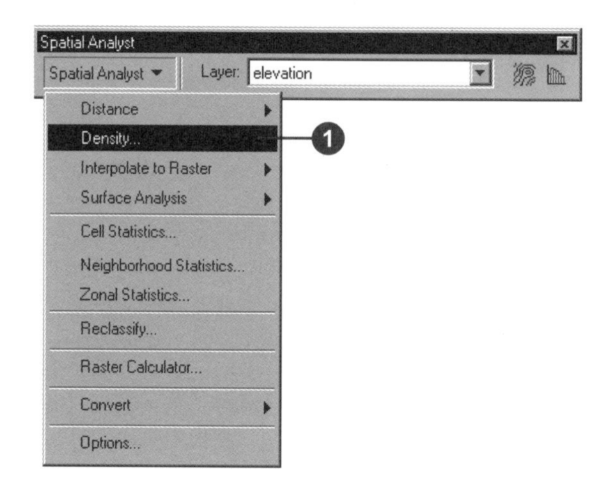

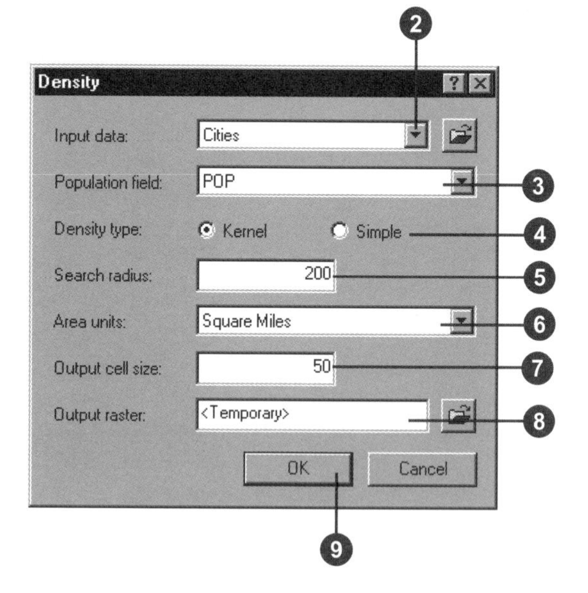

Interpolating to raster

What is interpolation?

Interpolation predicts values for cells in a raster from a limited number of sample data points. It can be used to predict unknown values for any geographic point data: elevation, rainfall, chemical concentrations, noise levels, and so on.

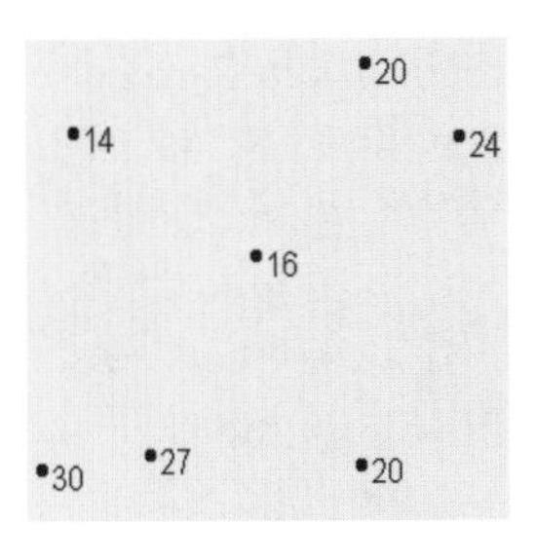

Point dataset of known values

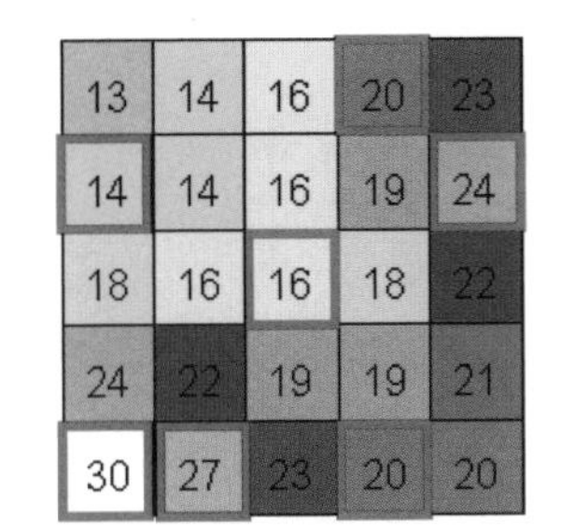

Raster Interpolated from the points. Cells highlighted in red indicate the values of the input point dataset.

The left-hand graphic above is a point dataset of known values. The right-hand graphic is a raster interpolated from these points. Unknown values are predicted with a mathematical formula that uses the values of nearby known points.

Why interpolate to raster?

Visiting every location in a study area to measure the height, magnitude, or concentration of a phenomenon is usually difficult or expensive. Instead, strategically dispersed sample input point locations can be selected and a predicted value can be assigned to all other locations. Input points can be either randomly or regularly spaced points containing height, concentration, or magnitude measurements.

The assumption that makes interpolation a viable option is that spatially distributed objects are spatially correlated; in other words, things that are close together tend to have similar characteristics. For instance, if it is raining on one side of the street, you can predict with a high level of confidence that it is also raining on the other side of the street. You would be less sure if it was raining across town and less confident still about the state of the weather in the next county. Using this analogy, it is easy to see that the values of points close to sampled points are more likely to be similar than those that are further apart. This is the basis of interpolation.

A typical use for point interpolation is to create an elevation surface from a set of sample measurements. Each symbol in the point layer represents a location where the elevation has been measured. By interpolating, the values between these input points will be predicted.

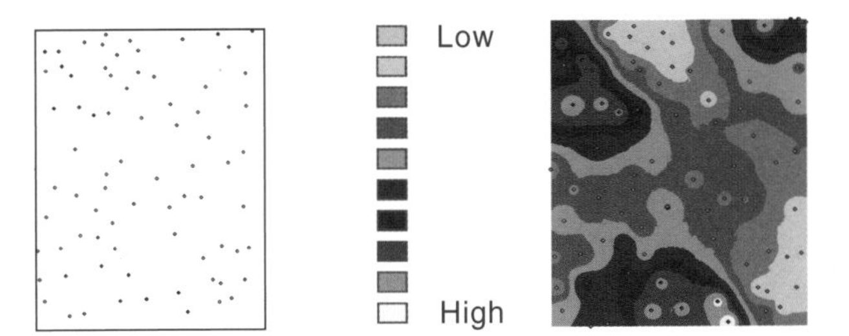

Details on the interpolators

The available interpolation methods are *Inverse Distance Weighted*, *Spline*, and *Kriging*. They make certain assumptions about how to determine the best estimated values. Based on the phenomena the values represent and on how the sample points are distributed, different interpolators will produce better estimates relative to the actual values. No matter which interpolator is selected, however, the more input points and the greater their distribution, the more reliable the results.

Inverse Distance Weighted

What is Inverse Distance Weighted (IDW)?

IDW estimates cell values by averaging the values of sample data points in the vicinity of each cell. The closer a point is to the center of the cell being estimated, the more influence, or weight, it has in the averaging process. This method assumes that the variable being mapped decreases in influence with distance from its sampled location. For example, when interpolating a surface of consumer purchasing power for a retail site analysis, the purchasing power of a more distant location will have less influence because people are more likely to shop closer to home.

Power

With IDW you can control the significance of known points upon the interpolated values, based upon their distance from the output point. By defining a high power, more emphasis is placed on the nearest points, and the resulting surface will have more detail—be less smooth. Specifying a lower power will give more influence to the points that are furthur away, resulting in a smoother surface. A power of 2 is most commonly used, and is the default.

Search radius

The characteristics of the interpolated surface can also be controlled by applying a search radius, fixed or variable, which limits the number of input points that can be used for calculating each interpolated cell.

Fixed search radius

A fixed search radius requires a distance and a minimum number of points. The distance dictates the radius of the circle of the neighborhood, in map units. The distance of the radius is constant, so for each interpolated cell, the radius of the circle used to find input points is the same. The minimum number of points indicates the minimum number of measured points to use within the neighborhood. All the measured points that fall within the radius will be used in the calculation of each interpolated cell. When there are fewer measured points in the neighborhood than the specified minimum, the search radius will increase until it can encompass the minimum number of points. The specified fixed search radius will be used for each interpolated cell—cell center—in the study area; thus, if your measured points are not spread out equally, which they rarely are, then there are likely to be a different number of measured points used in the different neighborhoods for the various predictions.

Variable search radius

With a variable search radius, the number of points used in calculating the value of the interpolated cell is specified, which makes the radius distance vary for each interpolated cell, depending on how far it has to search around each interpolated cell to reach the specified number of input points. Thus, some neighborhoods can be small and others can be large, depending on the density of the measured points near the interpolated cell. You can also specify a maximum distance, in map units, that the search radius cannot exceed. If the radius for a particular neighborhood reaches the maximum distance before obtaining the specified number of points, the prediction for that location will be performed on the number of measured points within the maximum distance.

Barrier

A *barrier* is a polyline dataset used as a break that limits the search for input sample points. A polyline can represent a cliff, ridge, or some other interruption in a landscape. Only those input sample points on the same side of the barrier as the current processing cell will be considered.

Inverse Distance Weighted interpolation

IDW has two options: a Fixed search radius type and a Variable search radius type.

With a Fixed radius, the radius of the circle used to find input points is the same for each interpolated cell. By specifying a minimum count, you can ensure that within the fixed radius, at least a minimum number of input points will be used in the calculation of each interpolated cell.

A higher Power puts more emphasis on the nearest points, creating a surface that has more detail but is less smooth. A lower Power gives more influence to surrounding points that are farther away, creating a smoother surface.

Use a barrier to limit the search for input sample points to the side of the barrier on which the interpolated cell sits, such as in the case of a cliff or a ridge.

With a variable radius, the count represents the number of points used in calculating the value of the interpolated cell. This makes the search radius variable for each interpolated cell, depending on how far it ▸

Creating a surface using IDW with a Fixed radius

1. Click the Spatial Analyst dropdown arrow, point to Interpolate to Raster, and click Inverse Distance Weighted.
2. Click the Input points dropdown arrow and click the point dataset you wish to use.
3. Click the Z value field dropdown arrow and click the field you wish to use.
4. Optionally, change the default Power value.
5. Click the Search radius type dropdown arrow and click Fixed.
6. Optionally, change the default Distance for the search radius. The default radius is five times the cell size of the output raster.
7. Optionally, change the Minimum number of points.
8. Optionally, specify a barrier.
9. Optionally, change the default Output cell size.
10. Specify a name for the Output raster or leave the default to create a temporary dataset in your working directory.
11. Click OK.

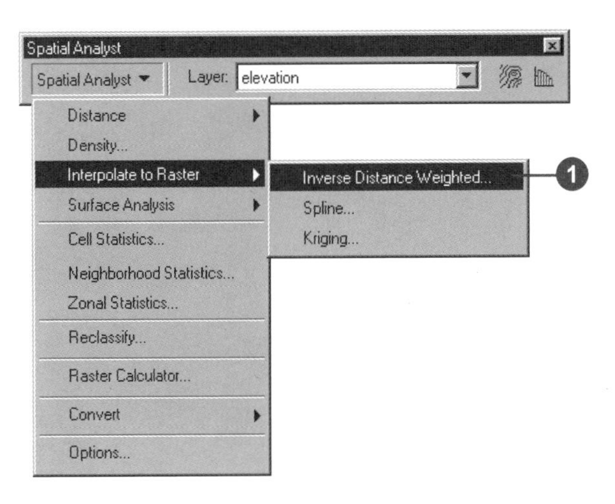

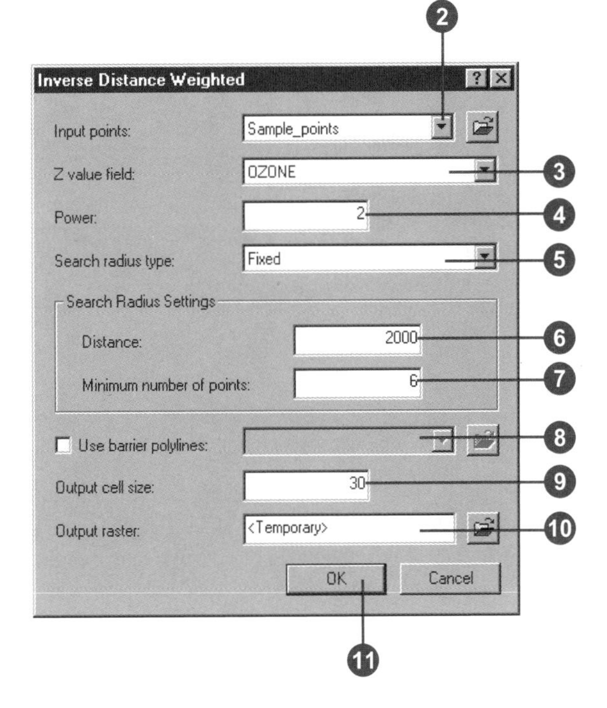

has to stretch to reach the specified number of input points.

Specify a maximum distance to limit the potential size of the radius of the circle. If the number of points is not reached before the maximum distance of the radius is reached, fewer points will be used in the calculation of the interpolated point.

Tip

Deciding on the distance or the number of points

Use the Measure tool on the Tools toolbar to measure distance between points to get an idea of the distance and number of points to use when setting the Search Radius.

Tip

Fixed or variable?

Use a fixed radius type if your input sample points are plentiful and are regularly spaced. Use a variable radius type if your sample points are sparse and are randomly placed.

Creating a surface using IDW with a variable radius

1. Click the Spatial Analyst dropdown arrow, point to Interpolate to Raster, and click Inverse Distance Weighted.
2. Click the Input points dropdown arrow and click the point dataset you wish to use.
3. Click the Z value field dropdown arrow and click the field you wish to use.
4. Optionally, change the default Power value.
5. Click the Search radius type dropdown arrow and click Variable.
6. Optionally, change the default number of points to use in the calculation of each interpolated point.
7. Specify a maximum distance for the radius to expand to in search of the number of points specified.
8. Optionally, specify a barrier.
9. Optionally, change the default Output cell size.
10. Specify a name for the Output raster or leave the default to create a temporary dataset in your working directory.
11. Click OK.

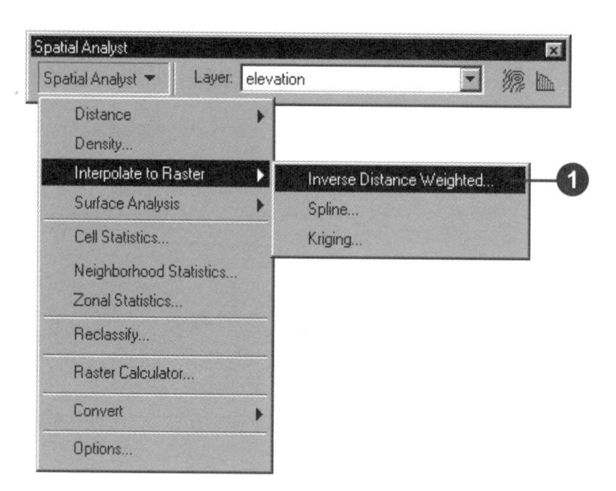

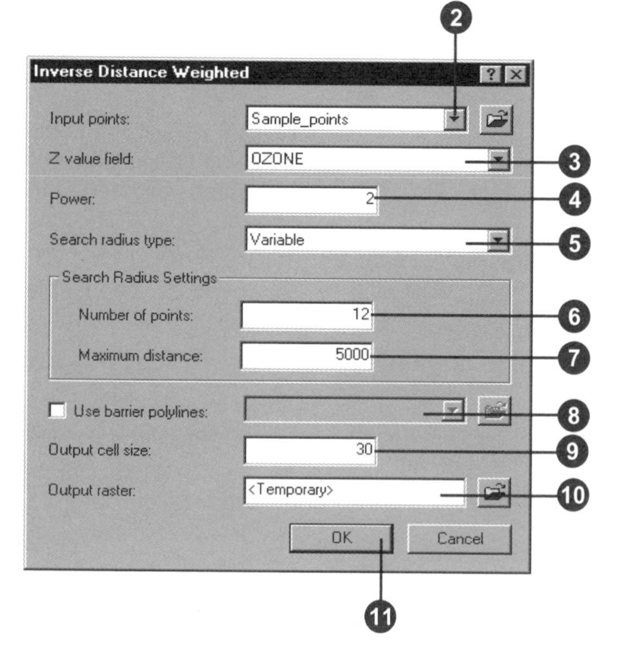

Spline

What is Spline?

Spline estimates values using a mathematical function that minimizes overall surface curvature, resulting in a smooth surface that passes exactly through the input points.

Conceptually, it is like bending a sheet of rubber to pass through the points while minimizing the total curvature of the surface. It fits a mathematical function to a specified number of nearest input points while passing through the sample points. This method is best for gently varying surfaces, such as elevation, water table heights, or pollution concentrations.

Spline methods

There are two Spline methods: regularized and tension.

Regularized

The Regularized method creates a smooth, gradually changing surface with values that may lie outside the sample data range.

Tension

The Tension method tunes the stiffness of the surface according to the character of the modeled phenomenon. It creates a less-smooth surface with values more closely constrained by the sample data range.

Optional parameters

Weight

For Regularized, weight defines the weight of the third derivatives of the surface in the curvature minimization expression. The higher the weight, the smoother the surface. The values entered for this parameter must be equal to or greater than zero. The typical values that may be used are 0, .001, .01, .1, and .5.

For Tension, weight defines the weight of tension. The higher the weight, the coarser the surface. The values entered have to be equal to or greater than zero. The typical values are 0, 1, 5, and 10.

Number of points

Number of points identifies the number of points used in the calculation of each interpolated cell. The more input points you specify, the more each cell is influenced by distant points and the smoother the surface.

Spline interpolation

The Regularized Spline type ensures that you create a smooth surface and slope.

The Tension Spline type tunes the stiffness of the surface according to the character of the modeled phenomenon.

The Number of points option specifies the number of points used in the calculation of each interpolated point. The more input points you specify, the smoother the surface.

Tip

Choosing a weight for Spline interpolations

Regularized spline: *The higher the weight, the smoother the surface. Weights between 0 and 5 are suitable. Typical values are 0, .001, .01, .1, and .5.*

Tension spline: *The higher the weight, the coarser the surface and the more the values conform to the range of sample data. Weight values must be greater than or equal to zero. Typical values are 0, 1, 5, and 10.*

Creating a surface using Spline interpolation

1. Click the Spatial Analyst dropdown arrow, point to Interpolate to Raster, and click Spline.
2. Click the Input points dropdown arrow and click the point dataset you wish to use.
3. Click the Z value field dropdown arrow and click the field you wish to use.
4. Click the Spline type dropdown arrow and click the Spline method you wish to use.
5. Optionally, change the default Weight.

   For the Regularized method, the higher the weight, the smoother the surface. For the Tension method, the higher the weight, the coarser the surface.
6. Optionally, change the default Number of points to use in the calculation of each interpolated point.
7. Optionally, change the default Output cell size.
8. Specify a name for the Output raster or leave the default to create a temporary dataset in your working directory.
9. Click OK.

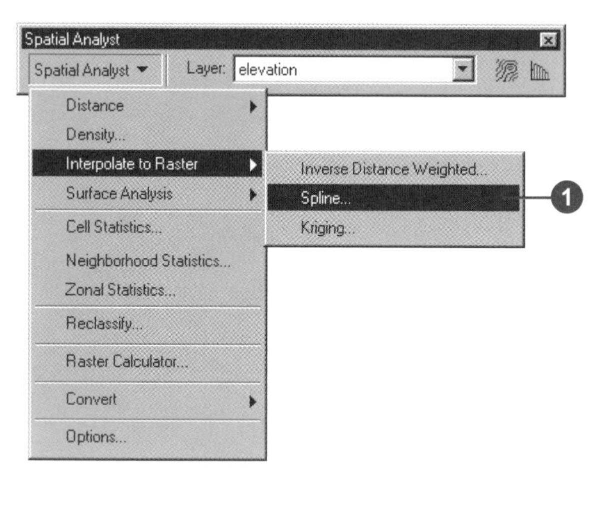

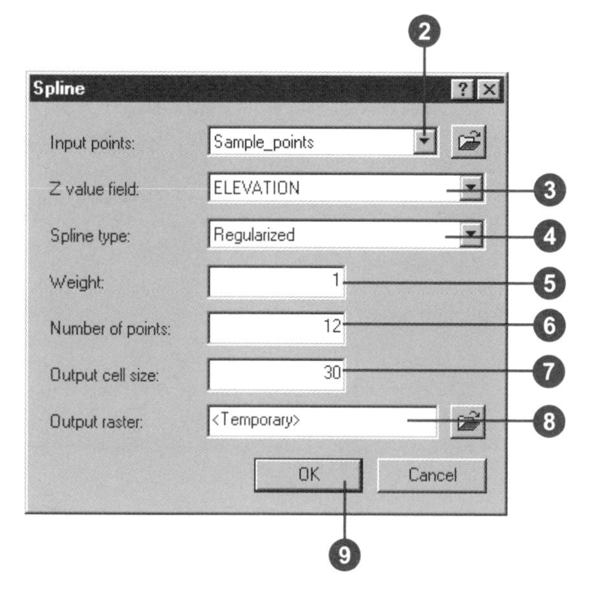

Kriging

What is Kriging?

IDW and Spline, discussed earlier, are referred to as deterministic interpolation methods because they are directly based on the surrounding measured values or on specified mathematical formulas that determine the smoothness of the resulting surface. A second family of interpolation methods consists of geostatistical methods, such as Kriging, that are based on statistical models that include autocorrelation—the statistical relationship among the measured points. Because of this, not only do these techniques have the capability of producing a prediction surface, but they can also provide some measure of the certainty or accuracy of the predictions.

Kriging is similar to IDW in that it weights the surrounding measured values to derive a prediction for an unmeasured location. The general formula for both interpolators is formed as a weighted sum of the data:

$$\hat{Z}(\mathbf{s}_0) = \sum_{i=1}^{N} \lambda_i Z(\mathbf{s}_i)$$

where

$Z(\mathbf{s}_i)$ is the measured value at the ith location;

λ_i is an unknown weight for the measured value at the ith location;

$\mathbf{s}_0$ is the prediction location;

N is the number of measured values.

In IDW, the weight, λ_i, depends solely on the distance to the prediction location. However, in Kriging, the weights are based not only on the distance between the measured points and the prediction location, but also on the overall spatial arrangement among the measured points and their values. To use the spatial arrangement in the weights, the spatial autocorrelation must be quantified. Thus, in Ordinary Kriging, the weight, λ_i, depends on a fitted model to the measured points, the distance to the prediction location, and the spatial relationships among the measured values around the prediction location.

To make a prediction with Kriging, two tasks are necessary: (1) to uncover the dependency rules and (2) to make the predictions. To realize these two tasks, Kriging goes through a two-step process: (1) variograms and covariance functions are created to estimate the statistical dependence—called spatial autocorrelation—values, which depends on the model of autocorrelation—fitting a model—and (2) prediction of unknown values are made. It is because of these two distinct tasks that it has been said that Kriging uses the data twice: the first time to estimate the spatial autocorrelation of the data and the second time to make the predictions.

Variography

Fitting a model, or spatial modeling, is also known as structural analysis, or variography. In spatial modeling of the structure of the measured points, we begin with a graph of the empirical semivariogram, computed as:

Semivariogram(distance h) = 0.5 * average[(value at location i – value at location j)^2]

for all pairs of locations separated by distance h. The formula involves calculating the difference squared between the values of the paired locations. The diagram that follows shows the pairing of one point—the red point—with all other measured locations. This process continues for each measured point.

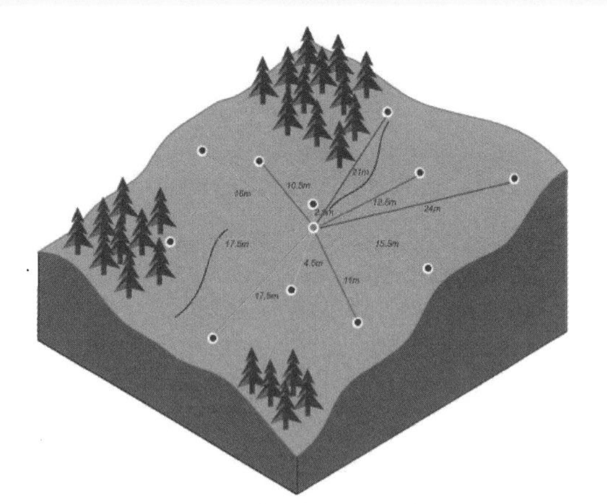

The pairing of one point—the red point—with all other measured locations

Often, each pair of locations has a unique distance, and there are often many pairs of points. To plot all pairs quickly becomes unmanageable. Instead of plotting each pair, the pairs are grouped into *lag bins*. For example, compute the average semivariance for all pairs of points that are greater than 40 meters apart but less than 50 meters. The empirical semivariogram is a graph of the averaged semivariogram values on the y-axis and distance, or lag, on the x-axis (see diagram below).

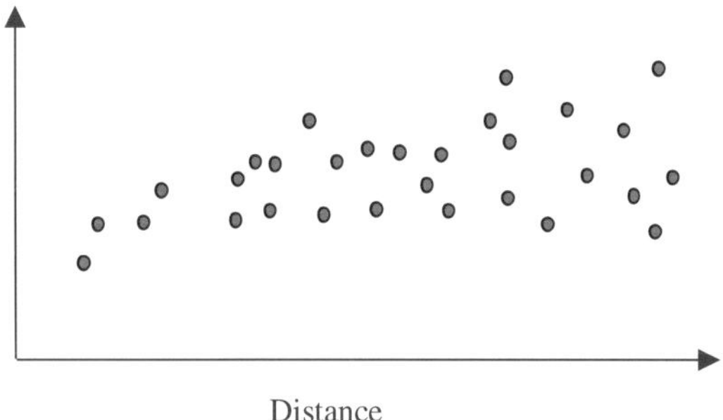

Spatial autocorrelation quantifies a basic principle of geography—things that are closer are more alike than things farther apart. Thus, pairs of locations that are closer—far left on the x-axis of the semivariogram cloud—should have more similar values—low on the y-axis of the semivariogram cloud. As pairs of locations become farther apart—moving to the right on the x-axis of the semivariogram cloud—they should become more dissimilar and have a higher squared difference—move up on the y-axis of the semivariogram cloud.

Fitting a model to the empirical semivariogram

The next step is to fit a model to the points forming the empirical semivariogram. Semivariogram modeling is a key step between spatial description and spatial prediction. The main application of Kriging is the prediction of attribute values at unsampled locations. We have seen how the empirical semivariogram provides information on the spatial autocorrelation of datasets. However, it does not provide information for all possible directions and distances. For this reason, and to ensure that Kriging predictions have positive Kriging variances, it is necessary to fit a model—that is, a continuous function or curve—to the empirical semivariogram. Abstractly, this is similar to regression analysis, where a continuous line or curve is fitted.

We select some function that serves as our model—for example, a spherical type that rises at first and then levels off for larger distances beyond a certain range. There are deviations of the points on the empirical semivariogram from the model; some points are above the model curve, and some points are below. But, if we add the distance each point is above the line and add the distance each point is below the line, the two values should be similar. There are a lot of different semivariogram models to choose from.

Different types of semivariogram models

Spatial Analyst provides the following functions to choose from to model the empirical semivariogram: Circular, Spherical, Exponential, Gaussian, and Linear. The selected model influences the prediction of the unknown values, particularly when the shape of the curve near the origin differs significantly. The steeper the curve is near the origin, the more influence the closest neighbors will have on the prediction. As a result, the output surface will be less smooth. Each model is designed to fit different types of phenomenon more accurately.

The diagrams below show two common models and identify how the functions differ:

- The Spherical model

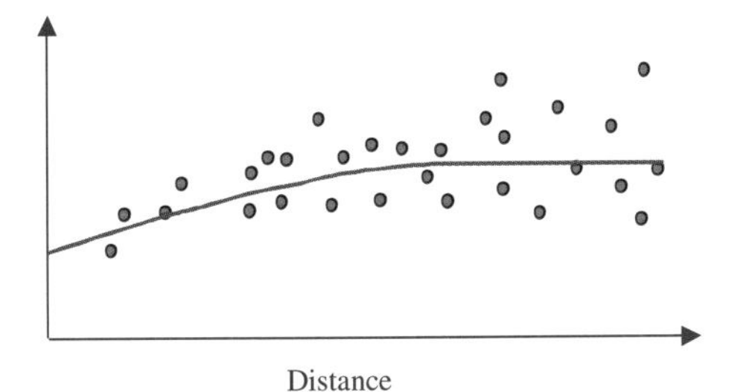

This model shows a progressive decrease of spatial autocorrelation—equivalently, an increase of semivariance—until some distance, beyond which autocorrelation is zero. The spherical model is one of the most commonly used models.

- The Exponential model

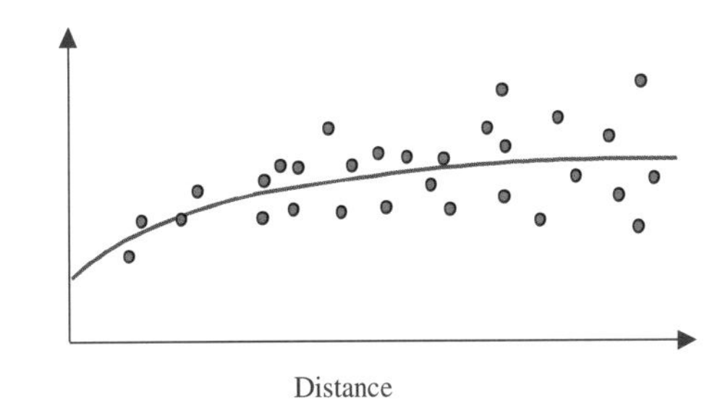

This model is applied when spatial autocorrelation decreases exponentially with increasing distance. Here, the autocorrelation disappears completely only at an infinite distance. The exponential model is also a commonly used model.

The choice of which model to use in Spatial Analyst is based on the spatial autocorrelation of the data and on prior knowledge of the phenomenon.

Understanding a semivariogram—the range, sill, and nugget

As previously discussed, the semivariogram depicts the spatial autocorrelation of the measured sample points. Because of a basic principle of geography—things that are closer are more alike—measured points that are closer will generally have a smaller difference squared than those farther apart. Once each pair of locations is plotted, after being binned, a model is fit through them. There are certain characteristics that are commonly used to describe these models.

The range and sill

When you look at the model of a semivariogram, you will notice that at a certain distance the model levels out. The distance where the model first flattens out is known as the range. Sample locations separated by distances closer than the range are spatially autocorrelated, whereas locations farther apart than the range are not.

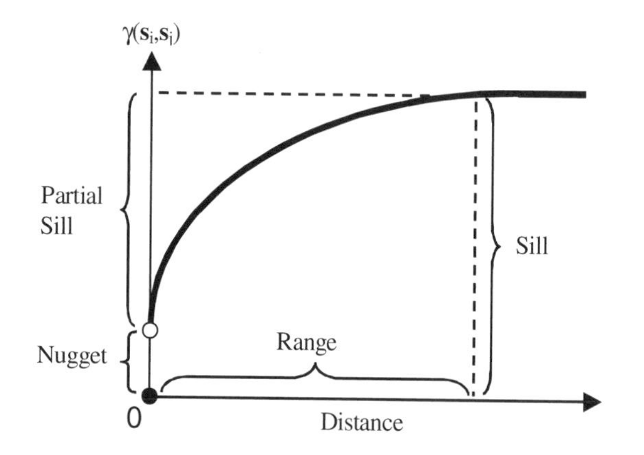

The value at which the semivariogram model attains the range—the value on the y-axis—is called the sill. The partial sill is the sill minus the nugget (see following section).

The nugget

Theoretically, at zero separation distance—that is, lag = 0—the semivariogram value is zero. However, at an infinitely small separation distance, the semivariogram often exhibits a nugget effect, which is some value greater than zero. If the semivariogram model intercepts the y-axis at 2, then the nugget is 2.

The nugget effect can be attributed to measurement errors or spatial sources of variation at distances smaller than the sampling interval, or both. Measurement error occurs because of the error inherent in measuring devices. Natural phenomema can vary spatially over a range of scales—that is, micro or macro scales. Variation at micro scales smaller than the sampling distances will appear as part of the nugget effect. Before collecting data, it is important to gain some understanding of the scales of spatial variation that you are interested in.

Making a prediction

The first task of uncovering the dependence, autocorrelation, in your data has been accomplished. You have also finished with the first use of the data, where the spatial information in the data, to compute distances, is used to model the spatial autocorrelation. Once you have the spatial autocorrelation, proceed with prediction using the fitted model; thereafter, the empirical semivariogram is set aside.

For the second task, use the data again to make predictions. Like IDW interpolation, Kriging forms weights from surrounding measured values to predict at unmeasured locations. As with IDW interpolation, the measured values closest to the unmeasured locations have the most influence. However, the Kriging weights for the surrounding measured points are more sophisticated than those of IDW. IDW uses a simple algorithm based on distance, but Kriging weights come from a semivariogram model that was developed by looking at the spatial nature of the data. To create a continuous surface or map of the phenomenon, predictions are made for each location—cell centers—in the study area based on the semivariogram model and the spatial arrangement of measured values that are nearby.

Search radius

We know from a basic principle of geography that things that are close to one another are more alike than things farther away. Using this principle, we can assume that as the locations get farther from the prediction location, the measured values will have less spatial autocorrelation with the unknown value for the location we are predicting. Thus, we can eliminate those farther locations with little influence. Not only is there less relationship with farther locations, it is possible that the farther locations may have a negative influence if they are located in an area much different than the prediction location. Another reason to use search neighborhoods is for computational speed. The smaller the search neighborhood, the faster the predictions can be made. As a result, it is common practice to limit the number of points that are used when making a prediction by specifying a search neighborhood. The specified shape of the neighborhood restricts how far, and where, to look for the measured values to be used in each prediction. Other neighborhood parameters restrict the locations that will be used within that shape, such as defining the maximum and minimum number of measured points to use within the neighborhood.

You can determine the weights for the measured locations using the configuration of the valid points within the specified neighborhood around the prediction location in conjunction with the model fit to the semivariogram. From the weights and the values, a prediction can be made for the unknown value at the prediction location.

Spatial Analyst has two neighborhood types: fixed and variable.

Fixed search radius

A fixed search radius requires a distance and a minimum number of points. The distance dictates the radius of the circle of the neighborhood, in map units. The distance of the radius is constant, so for each interpolated cell, the radius of the circle used to find input points is the same. The Minimum number of points indicates the minimum number of measured points to use within the neighborhood. All the measured points that fall within the radius will be used in the calculation of each interpolated cell. When there are fewer measured points in the neighborhood than the specified minimum, the search radius will increase until it can encompass the minimum number of points. The specified fixed search radius will be used for each interpolated cell—cell center—in the study area; thus, if your measured points are not spread out equally, which they rarely are, then there will likely be a different number of measured points used in the different neighborhoods for the various predictions.

Variable search radius

With a variable search radius, the number of points used in calculating the value of the interpolated cell is specified, which makes the radius distance vary for each interpolated cell, depending on how far it has to search around each interpolated cell to reach the specified number of input points. Thus, some neighborhoods can be small and others can be large, depending on the density of the measured points near the interpolated cell. You can also specify a maximum distance, in map units, that the search radius cannot exceed. If the radius for a particular neighborhood reaches the maximum radius before obtaining the specified number of points, the prediction for that location will be performed on the number of measured points within the maximum radius.

Kriging methods

Spatial Analyst provides two kriging methods: Ordinary and Universal.

Ordinary kriging

Ordinary kriging is the most general and widely used of the kriging methods. It assumes the constant mean is unknown. This is a reasonable assumption unless there is some scientific reason to reject this assumption.

Universal kriging

Universal kriging assumes that there is an overriding trend in the data—for example, a prevailing wind—and it can be modeled by a deterministic function, or polynomial. This polynomial is subtracted from the original measured points, and the autocorrelation is modeled from the random errors. Once the model is fit to the random errors, before making a prediction, the polynomial is added back to the predictions to give you meaningful results. Universal kriging should only be used when you know there is a trend in your data and you can give a scientific justification to describe it.

Kriging interpolation

There are two kriging methods: Ordinary and Universal.

Ordinary kriging is the most general and widely used of the kriging methods and is the default. It assumes the constant mean is unknown. Universal kriging should only be used when you know there is a trend in your data and you can give a scientific justification to describe it.

By using a variable search radius, you can specify the number of points to use in calculating the value of the interpolated cell. This makes the search radius variable for each interpolated cell, depending on how far it has to stretch to reach the specified number of input points.

Specifying a maximum distance limits the potential size of the radius of the circle. If the number of points is not reached before the maximum distance of the radius is reached, fewer points will be used in the calculation of the interpolated cell. ▸

Creating a surface using kriging interpolation with a variable radius

1. Click the Spatial Analyst dropdown arrow, point to Interpolate to Raster, and click Kriging.
2. Click the Input points dropdown arrow and click the point dataset you wish to use.
3. Click the Z value field dropdown arrow and click the field you wish to use.
4. Click the Kriging method you wish to use.
5. Click the Semivariogram model dropdown arrow and click the model you wish to use.
6. Click the Search radius type dropdown arrow and click Variable.
7. Optionally, change the default number of points.
8. Optionally, specify a maximum distance.
9. Optionally, change the default Output cell size.
10. Optionally, check Create Prediction of standard error.
11. Specify a name for the Output raster or leave the default to create a temporary dataset in your working directory.
12. Click OK.

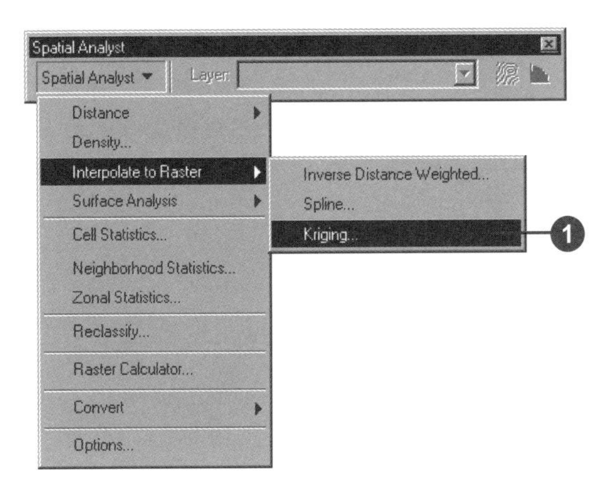

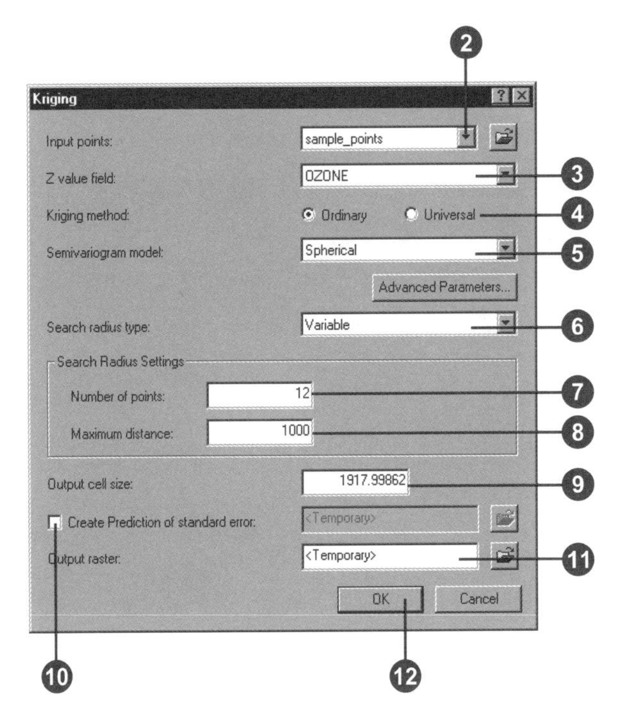

With a Fixed radius, the radius of the circle used to find input points is the same for each interpolated cell. The default radius is five times the cell size of the output raster. By specifying a minimum number of points, you can ensure that within the Fixed radius, at least a minimum number of input points will be used in the calculation of each interpolated cell.

Tip

Deciding on the radius or the number of points

Use the Measure tool on the Tools toolbar to measure distance between points to get an idea of the radius and number of points to use.

Tip

Changing the lag size, major range, partial sill, and nugget

Click Advanced Parameters on the Kriging dialog box to specify these parameters if they are known; otherwise, Spatial Analyst will estimate them for you.

Creating a surface using kriging interpolation with a fixed radius

1. Click the Spatial Analyst dropdown arrow, point to Interpolate to Raster, and click Kriging.
2. Click the Input points dropdown arrow and click the point dataset you wish to use.
3. Click the Z value field dropdown arrow and click the field you wish to use.
4. Click the Kriging method you wish to use.
5. Click the Semivariogram model dropdown arrow and click the model you wish to use.
6. Click the Search radius type dropdown arrow and click Fixed.
7. Optionally, change the default distance for the search radius setting.
8. Optionally, change the minimum number of points.
9. Optionally, change the default Output cell size.
10. Optionally, check Create Prediction of standard error.
11. Specify a name for the Output raster or leave the default to create a temporary dataset in your working directory.
12. Click OK.

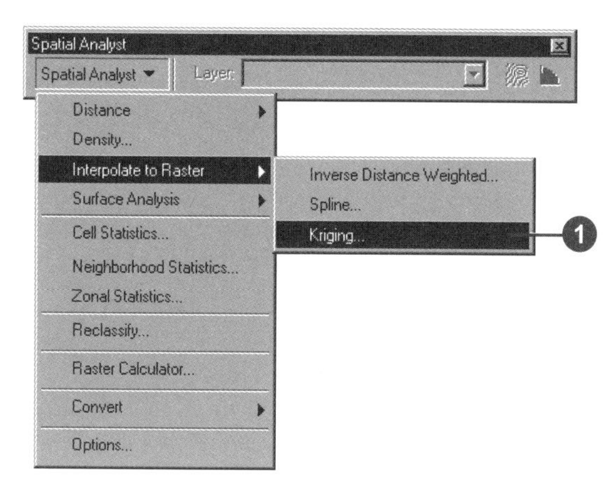

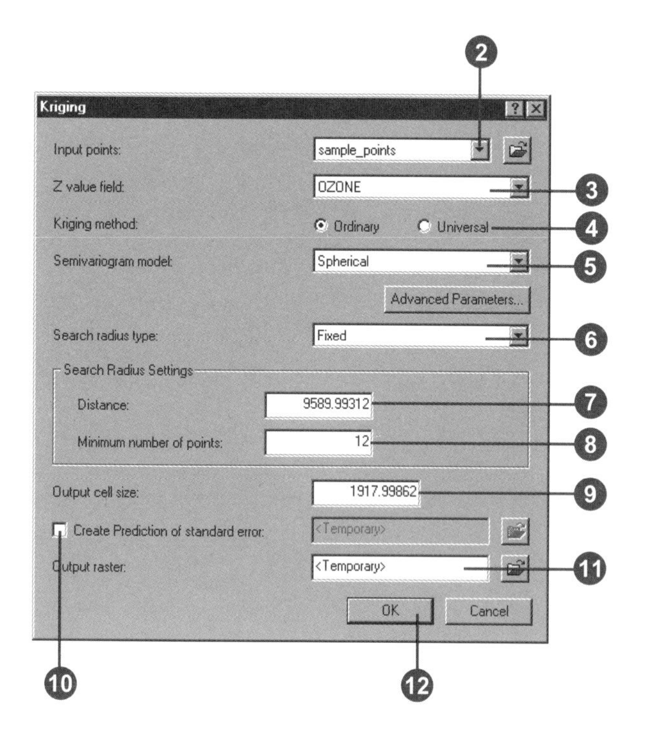

Performing surface analysis

You can gain additional information by producing a new dataset that identifies a specific pattern within an original dataset. Patterns that were not readily apparent in the original surface can be derived, such as contours, angle of slope, aspect, hillshade, viewshed, and cut/fill.

Contours can be useful for finding areas of the same value. You may be interested in obtaining elevation values for specific locations and examining the overall gradation of the land.

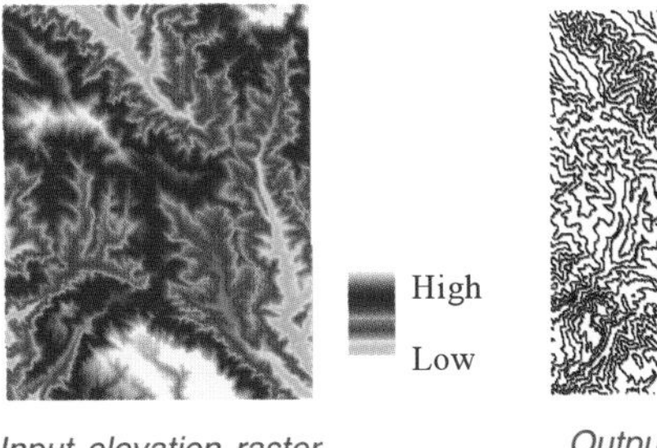

Input elevation raster

Output contours

You may, for instance, want to know the variations in the *slope* of the landscape because you want to find the areas most at risk of landslide based on the angle of slopes in an area—steeper slopes being those most at risk.

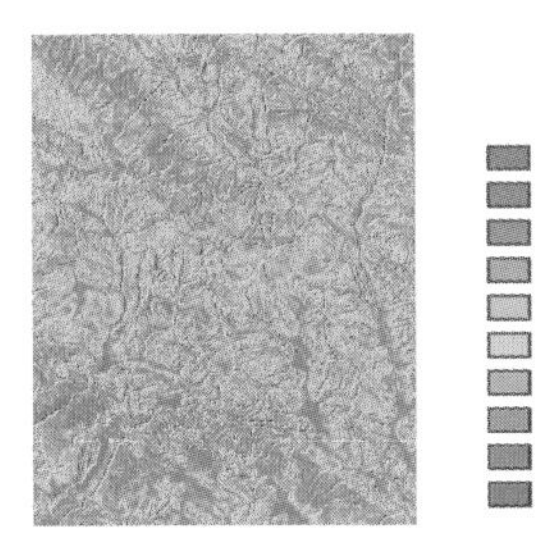

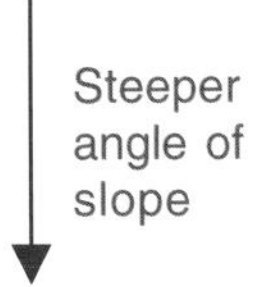

Output slope

You may be a farmer interested in locating a field on an area with a southerly *aspect*.

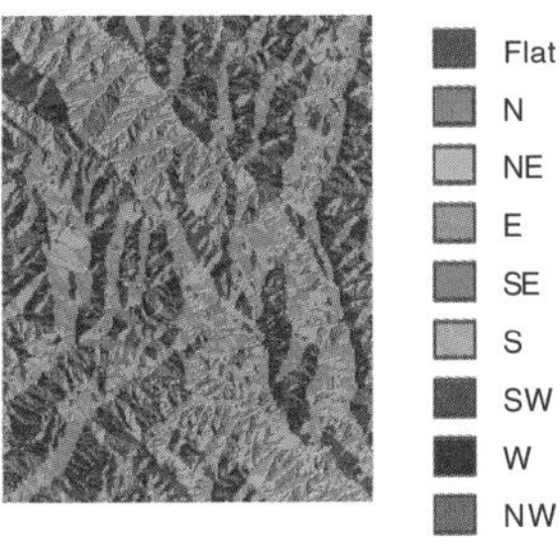

Output aspect

Output hillshade

You can create *a hillshade* for both analytical and graphical purposes. Graphically, a hillshade can provide an attractive and realistic backdrop showing how other layers are distributed in relation to the terrain relief.

From an analytical point of view, you can, for instance, analyze how the landscape is illuminated at various times of the day by lowering and raising the sun angle used in the analysis.

Azimuth 45°

Azimuth 315°

Calculating *viewshed* is useful when you want to know how visible objects will be. For instance, you might want to find the location with the most expansive view in an area because you want to know the best location for a lookout.

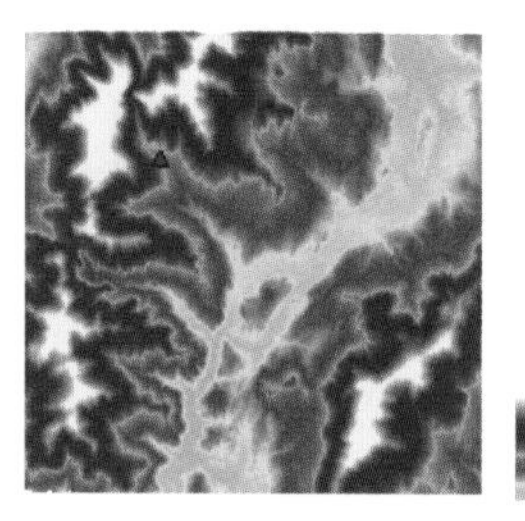

Input elevation raster

Output viewshed

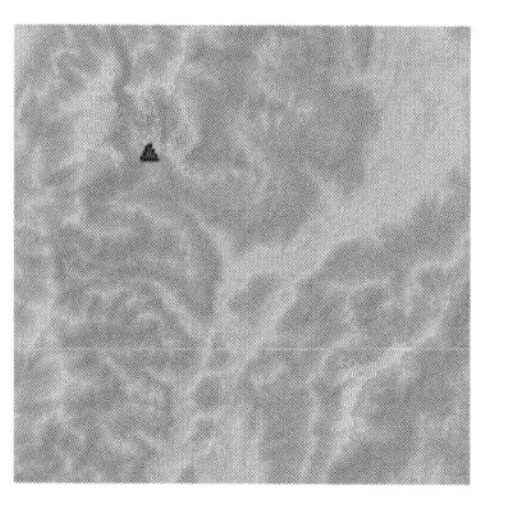

Display a hillshade transparently underneath the result from the Viewshed function.

It is useful to calculate a *Cut/Fill* surface when you want to know the areas and volumes of change between two surfaces. It identifies the areas and volumes of the surface that have been modified by the addition or removal of surface material. You may want to know the volumes and areas of surface material to be removed and filled in order to level a site for building construction, or you might want to identify areas that have been removed and areas that have been filled after a volcanic eruption.

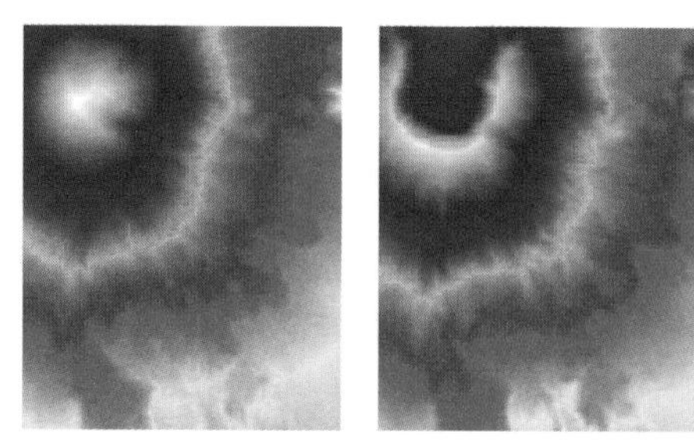

Before surface *After surface*

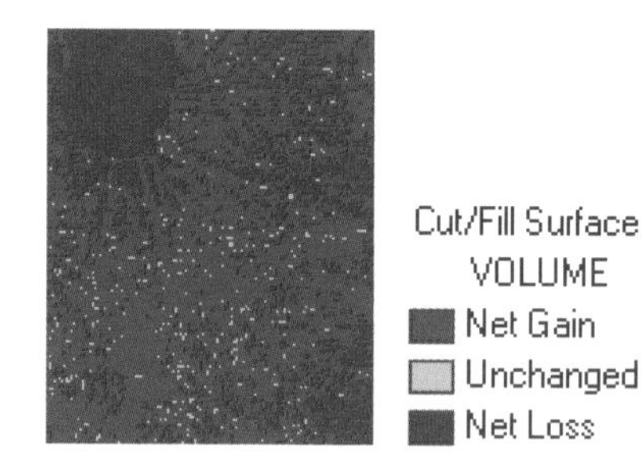

Cut/Fill surface

Contour

What are contours?

Contours are polylines that connect points of equal value, such as elevation, temperature, precipitation, pollution, or atmospheric pressure. The distribution of the polylines shows how values change across a surface. Where there is little change in a value, the polylines are spaced farther apart. Where the values rise or fall rapidly, the polylines are closer together.

Why create contours?

By following the polyline of a particular contour, you can identify which locations have the same value. Contours are also a useful surface representation because they allow you to simultaneously visualize flat and steep areas—distance between contours—and ridges and valleys—converging and diverging polylines.

The example below shows an input elevation dataset and the output contour dataset. The areas where the contours are closer together indicate the steeper locations. They correspond with the areas of higher elevation—in white on the input elevation dataset.

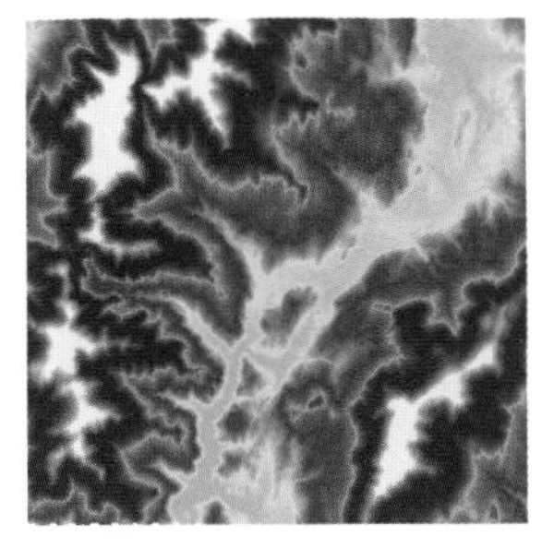

Input elevation dataset

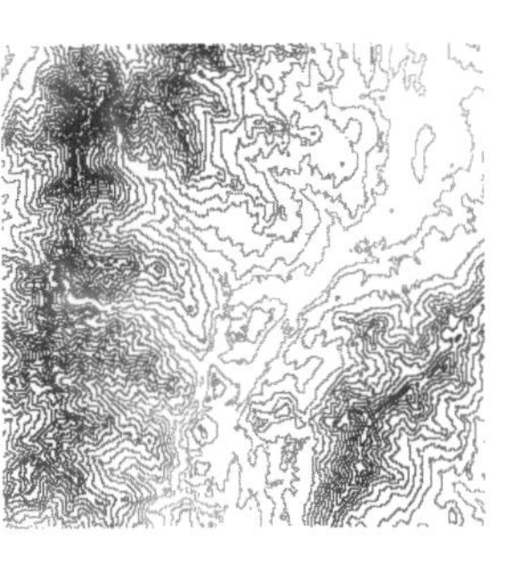

Output contour dataset

The Contour attribute table contains an elevation attribute for each contour polyline.

Attributes of Contour

FID	Shape	ID	CONTOUR
0	Polyline	1	1800
1	Polyline	2	1200
2	Polyline	3	1000
3	Polyline	4	2400
4	Polyline	5	1200
5	Polyline	6	3600
6	Polyline	7	1200
7	Polyline	8	1200
8	Polyline	9	1600
9	Polyline	10	3200

Record: 1 Show: All Selected R

Creating contours

The Contour function allows you to create contours for an entire dataset.

The Base contour is the value from which to begin generating contours. Contours are generated above and below this value as needed to cover the entire value range of the raster.

The Contour interval specifies the distance between contour polylines.

The *Z-factor* is the number of ground x,y units in one surface z unit. The Input surface values are multiplied by the specified Z-factor to adjust the Input surface z units to another measurement unit.

Tip

Using the Contour tool

Use the Contour tool on the Spatial Analyst toolbar to create contours for specific locations in your input dataset.

Tip

Highlighting contours

Use the Select Features tool on the Tools toolbar to select contours, then open the table to examine the values. Alternatively, select contours from the table.

Creating contours for your whole map

1. Click the Spatial Analyst dropdown arrow, point to Surface Analysis, and click Contour.
2. Click the Input surface dropdown arrow and click the surface you want to contour.
3. Type a Contour interval to specify the distance between contours.
4. Type a Base contour from which to start contouring, or leave the default of 0.
5. Optionally, type a value for the Z-factor.
6. Specify a name for the Output features or leave the default, which creates a permanent dataset in your working directory.
7. Click OK.

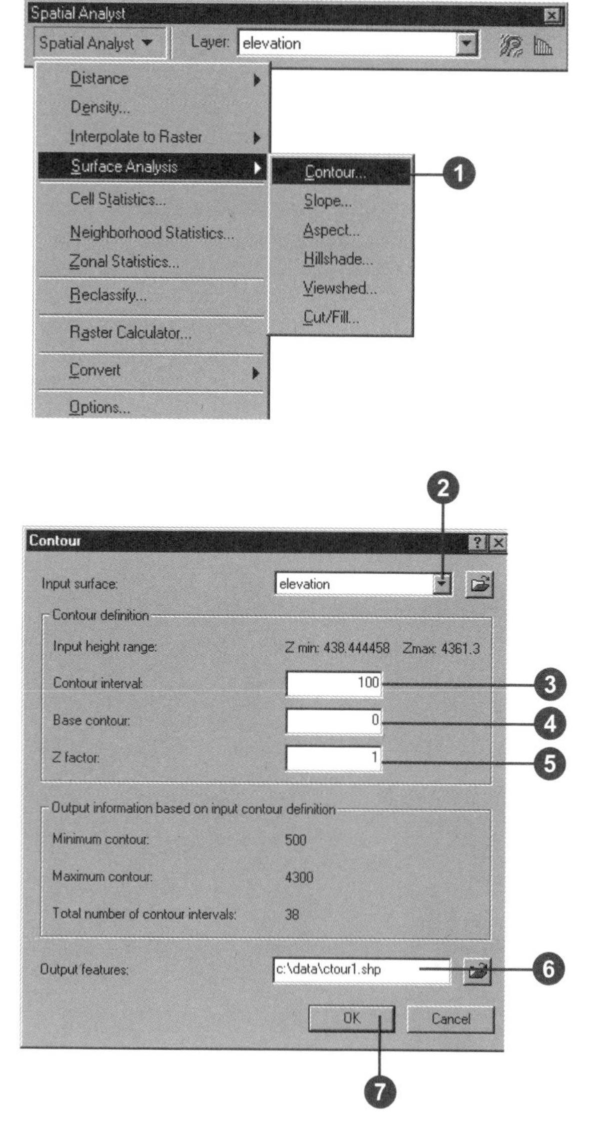

Slope

What is slope?

The Slope function calculates the maximum rate of change between each cell and its neighbors—the maximum change in elevation over distance between the cell and its eight neighbors—for example, the steepest downhill descent for the cell. Every cell in the output raster has a slope value. The lower the slope value, the flatter the terrain; the higher the slope value, the steeper the terrain. The output slope dataset can be calculated as percent slope or degree of slope.

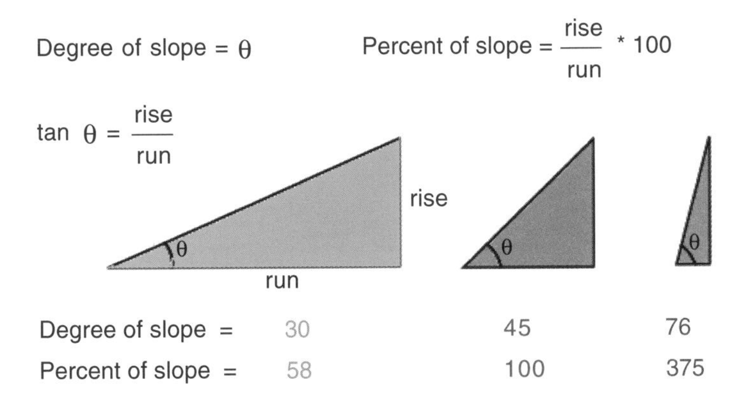

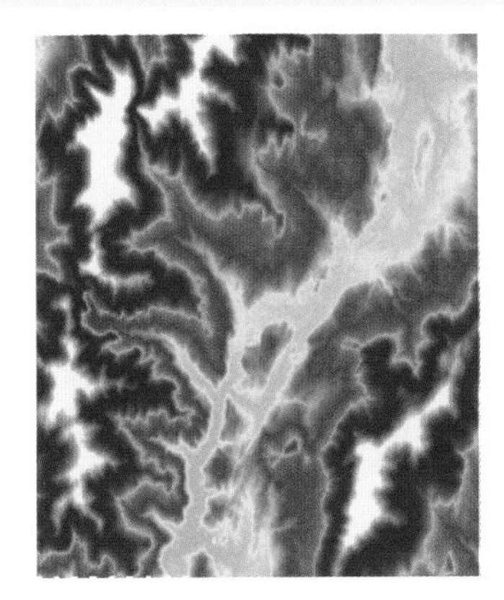

Elevation dataset

When the slope angle equals 45 degrees, the rise is equal to the run. Expressed as a percentage, the slope of this angle is 100 percent. Note that as the slope approaches vertical (90°), the percentage slope approaches infinity.

The Slope function is most frequently run on an elevation dataset, as the following diagrams show. Steeper slopes are shaded red on the output slope dataset. It can also be used with other types of continuous data, such as population, to identify sharp changes in value.

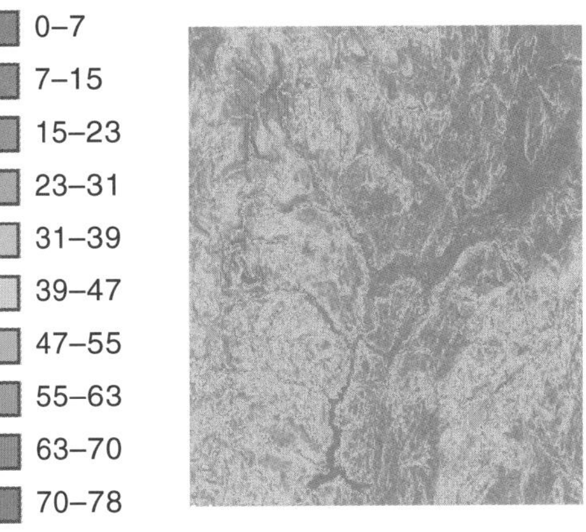

Output slope dataset (in degrees)

Calculating slope

The Slope function enables you to create a slope raster for an entire area, enabling you to get an impression of the steepness of the terrain and to use the output for further analysis.

The Z-factor is the number of ground x,y units in one surface z unit. The Input surface values are multiplied by the specified Z-factor to adjust the Input surface z units to another unit of measure.

Tip

Degree and percent slope

Slope can be measured in degrees from horizontal (0–90), or percent slope, which is the rise divided by the run, times 100.

Tip

Why use a Z-factor?

To get accurate slope results, the z units must be the same as the x,y units. If they are not the same, use a Z-factor to convert z units to x,y units. For example, if your x,y units are in meters, and your z units are in feet, you could use a Z-factor of 0.3048 to convert feet to meters.

Creating a slope dataset

1. Click the Spatial Analyst dropdown arrow, point to Surface Analysis, and click Slope.
2. Click the Input surface dropdown arrow and click the surface you want to calculate slope for.
3. Choose the Output measurement units.
4. Optionally, type a value for the Z-factor.
5. Optionally, change the default Output cell size.
6. Specify a name for the Output raster or leave the default to create a temporary dataset in your working directory.
7. Click OK.

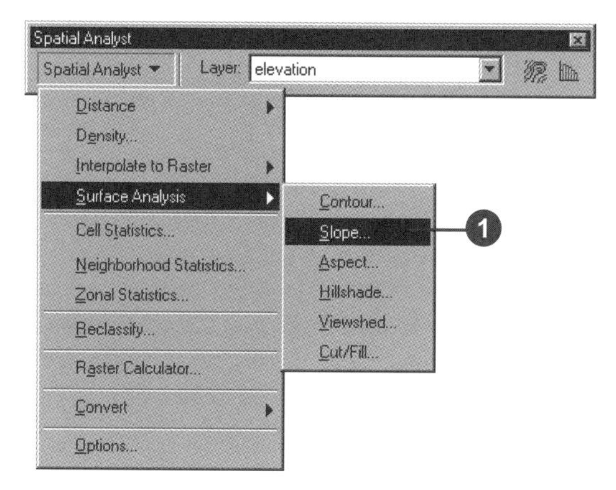

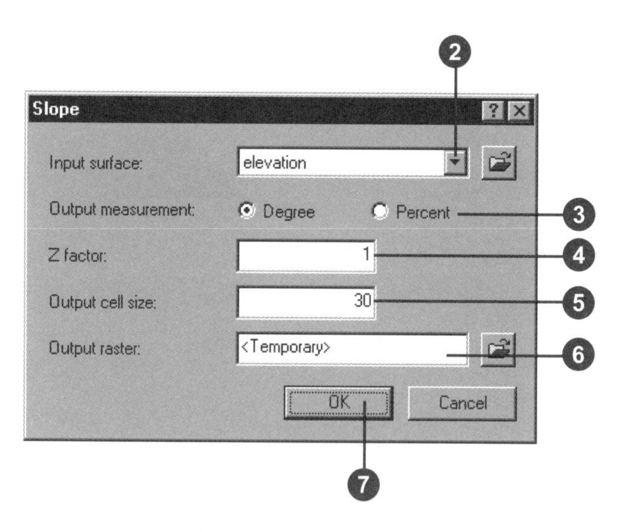

What is aspect?

Aspect identifies the steepest downslope direction from each cell to its neighbors. It can be thought of as slope direction or the compass direction a hill faces.

It is measured clockwise in degrees from 0—due north—to 360—again due north, coming full circle. The value of each cell in an aspect dataset indicates the direction the cell's slope faces. Flat slopes have no direction and are given a value of -1.

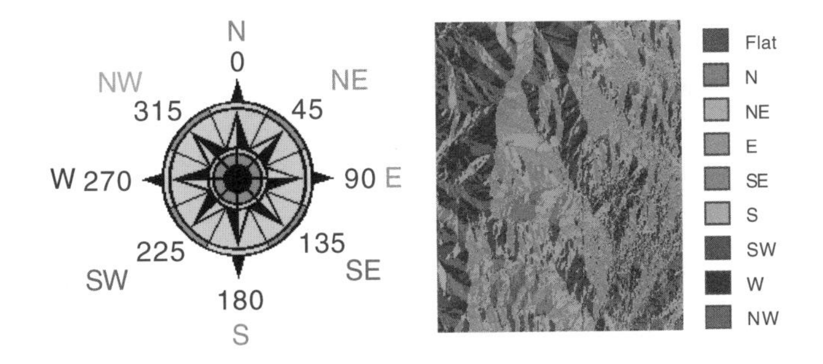

The diagram below shows an input elevation dataset and the output aspect raster.

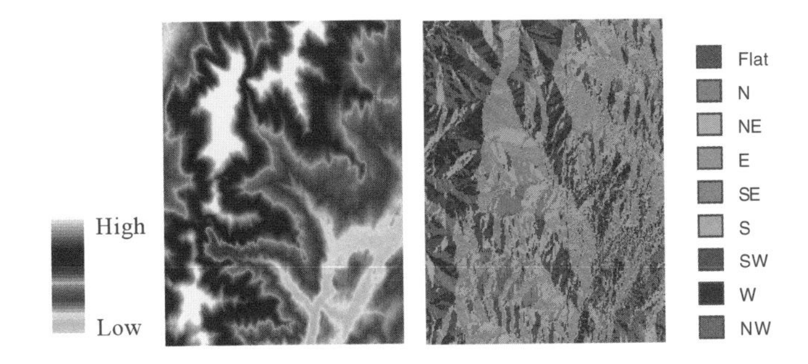

Why use the Aspect function?

With the Aspect function, you can:

- Find all north-facing slopes on a mountain as part of a search for the best slopes for ski runs.
- Calculate the solar illumination for each location in a region as part of a study to determine the diversity of life at each site.
- Find all southerly slopes in a mountainous region to identify locations where the snow is likely to melt first as part of a study to identify those residential locations that are likely to be hit by meltwater first.
- Identify areas of flat land to find an area for a plane to land in an emergency.

Calculating aspect

The Aspect function enables you to create a map displaying the steepest down-slope direction from each cell to its neighbors for an entire region. It is most commonly used with an elevation raster to identify the direction of slope.

Tip

Identifying slope direction

Use the Identify tool on the Tools toolbar to identify locations. This will give you the compass direction for a specific location on your output Aspect dataset.

Creating an aspect dataset

1. Click the Spatial Analyst dropdown arrow, point to Surface Analysis, and click Aspect.
2. Click the Input surface dropdown arrow and click the surface for which you want to calculate aspect.
3. Optionally, change the default Output cell size.
4. Specify a name for the Output raster or leave the default to create a temporary dataset in your working directory.
5. Click OK.

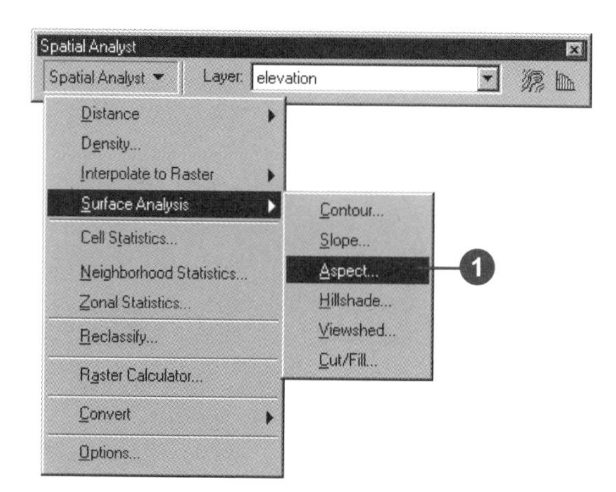

Hillshade

What is the Hillshade function?

The Hillshade function obtains the hypothetical illumination of a surface by determining illumination values for each cell in a raster. It does this by setting a position for a hypothetical light source and calculating the illumination values of each cell in relation to neighboring cells. It can greatly enhance the visualization of a surface for analysis or graphical display.

By default, shadow and light are shades of gray associated with integers from 0 to 255, increasing from black to white.

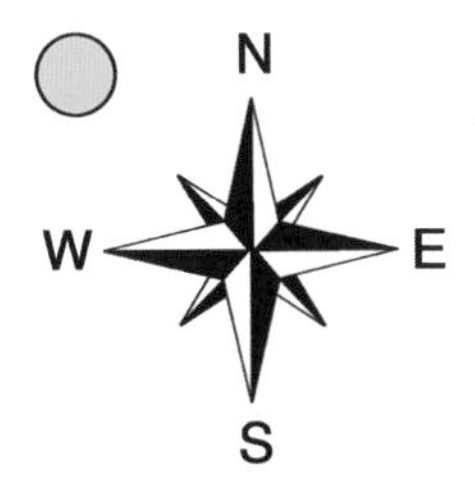

Azimuth is the angular direction of the sun, measured from north in clockwise degrees from 0 to 360. An azimuth of 90 is east. The default is 315 NW.

Altitude is the slope or angle of the illumination source above the horizon. The units are in degrees, from 0—on the horizon—to 90 degrees—overhead. The default is 45 degrees.

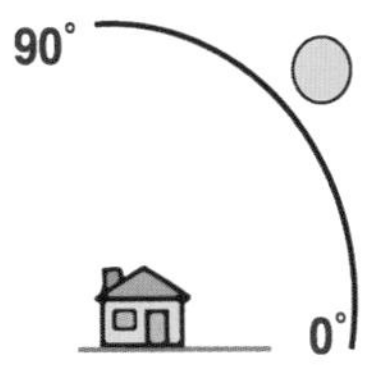

The hillshade to the left has an azimuth of 315 and an altitude of 45 degrees.

Using hillshading for display

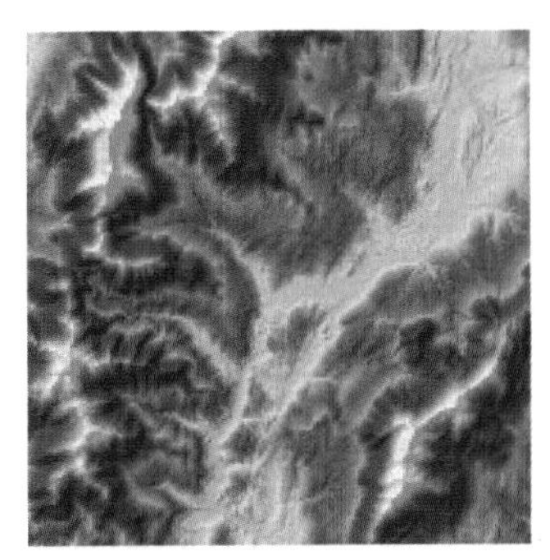

By placing an elevation raster on top of a created hillshade, then making the elevation raster transparent, you can create realistic images of the landscape. Add other layers, such as roads or streams, to further increase the informational content in the display.

Using hillshading in analysis

By modeling shade, the default option, you calculate the local illumination whether the cell falls in a shadow or not.

By modeling shadow, you can identify those cells that will be in the shadow of another cell at a particular time of day. Cells that are in the shadow of another cell are coded 0; all other cells are coded with integers from 1 to 255. You can reclassify values greater than 1 to 1, producing a binary output raster. In the example below, the black areas are in shadow. The azimuth is the same, but the sun angle—altitude—has been modified.

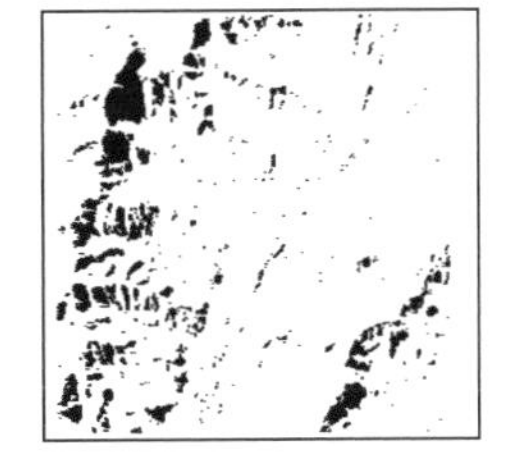

Sun angle: 45 degrees

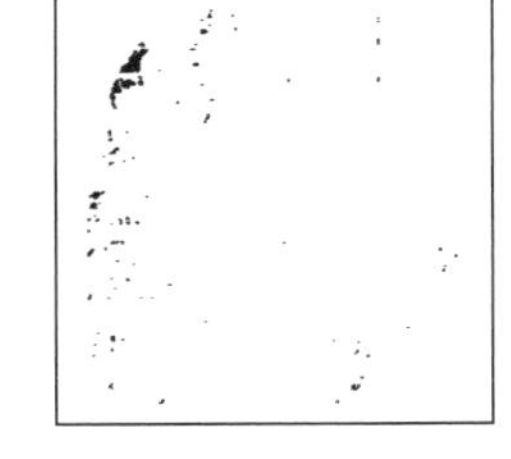

Sun angle: 60 degrees

Computing hillshade

The Hillshade function is typically used to create a shaded relief map from an elevation raster.

The default azimuth and altitude values work well for graphical display. For analysis, you may wish to modify these values.

Azimuth is the angular direction of the sun; the default angle of 315 is NW.

Altitude is the slope or angle of the illumination source above the horizon. The default is 45 degrees above the surface.

Tip

Why use a Z-factor?

To get accurate hillshade results, the z units must be the same as the x,y units. If they are not the same, use a Z-factor to convert z units to x,y units. For example, if your x,y units are in meters, and your z units are in feet, you could use a Z-factor of 0.3048 to convert feet to meters.

Tip

Modeling shadows

Checking Model shadows will assign a value of 0 to any cell that falls within a shadow. By accepting the default—not to model shadows—the local illumination will be calculated whether the cell falls in a shadow or not.

Creating a hillshade dataset

1. Click the Spatial Analyst dropdown arrow, point to Surface Analysis, and click Hillshade.
2. Click the Input surface dropdown arrow and click the surface for which you want to calculate hillshade.
3. Specify the azimuth you wish to use. The default is 315 degrees.
4. Specify an altitude. The default is 45 degrees.
5. Check Model shadows if you wish to model shadows by assigning a value of 0 to areas in shadow.

   Leaving this unchecked will create an output raster that will give local illumination regardless of shadows.
6. Specify a Z-factor. The default is 1.
7. Optionally, change the default Output cell size.
8. Specify a name for the Output raster or leave the default to create a temporary dataset in your working directory.
9. Click OK.

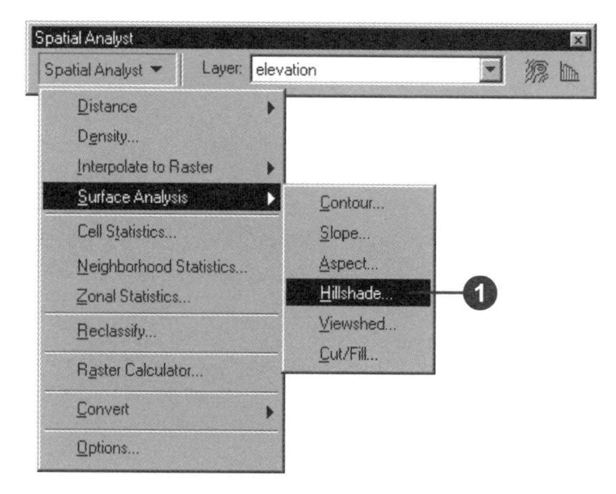

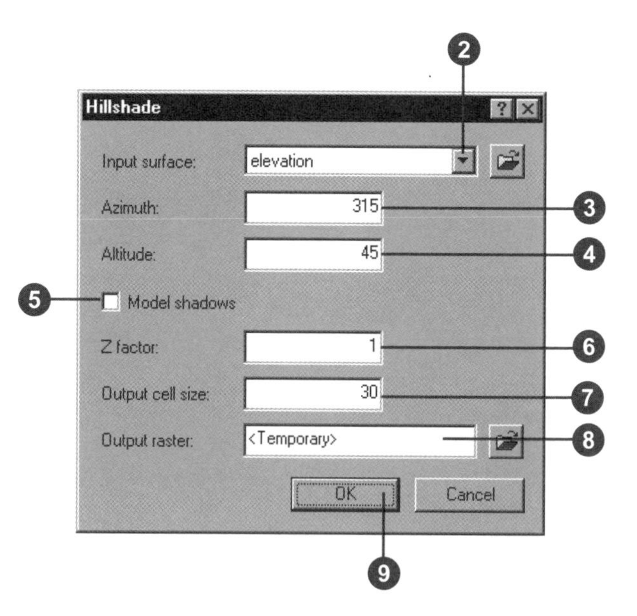

Using transparency

Transparency can be a useful tool for graphical display of your information. Applying a percentage of transparency to certain layers allows you to see multiple layers of information at the same time.

Transparency can be applied to both raster and feature data.

Tip

Adjusting brightness and contrast

Use the Adjust Brightness and Adjust Contrast buttons on the Effects toolbar to adjust the brightness or contrast of the hillshade layer for even better display.

Displaying hillshade transparently

1. Follow steps 1 through 9 for Creating a hillshade dataset.
2. Click and drag the elevation raster to the top of the table of contents, over the created hillshade.
3. Click View, point to Toolbars, and click Effects.
4. Click the Layer dropdown arrow and click elevation.
5. Click the Adjust Transparency button and move the scroll bar up to the desired level of transparency—try 30%.

   You should now see the hillshade underneath the elevation raster.

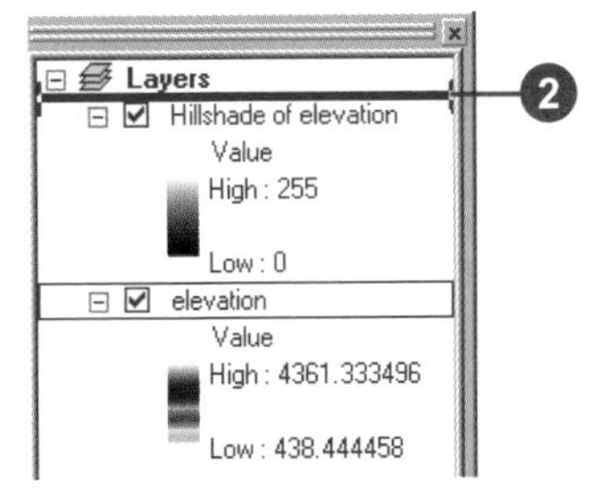

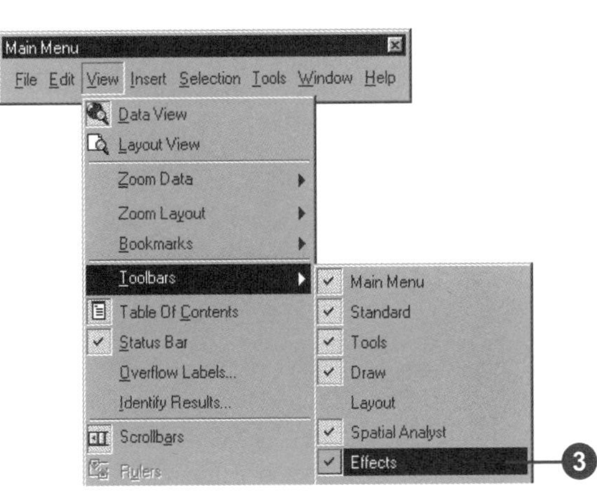

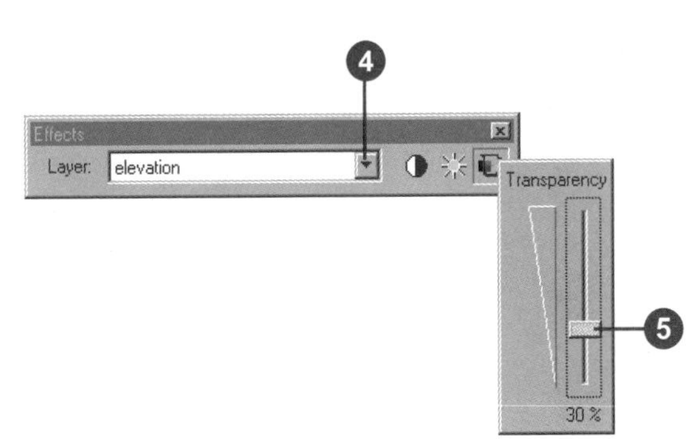

Viewshed

What is viewshed?

Viewshed identifies the cells in an input raster that can be seen from one or more observation points or lines. Each cell in the output raster receives a value that indicates how many observer points or lines can be seen from each location. If you have only one observer point, each cell that can see that observer point is given a value of 1. All cells that cannot see the observer point are given a value of 0.

The observer points feature class can contain points or lines. The nodes and vertices of lines will be used as observation points.

Why calculate viewshed?

Viewshed is useful when you want to know how visible objects might be, such as finding well-exposed places for communication towers.

In the example that follows, the viewshed from an observation point is identified. The elevation raster displays the height of the land—darker locations represent lower elevations—and the observation point is marked as a green triangle. Cells in green are visible from the observation point, and cells in red are not visible.

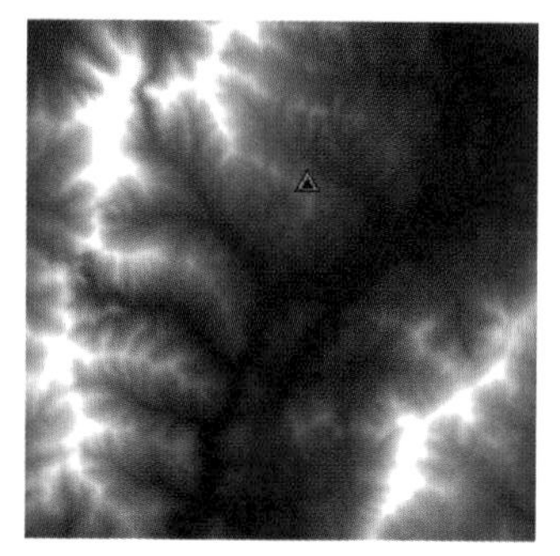

The elevation in the area of the observation point

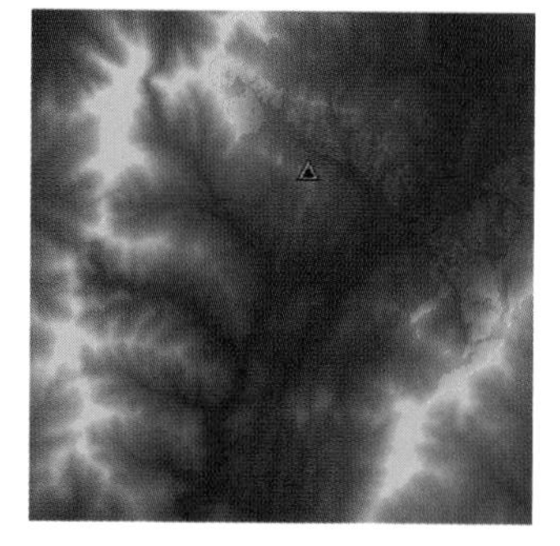

Green cells are visible from the observation point, red cells are not visible.

Displaying a hillshade underneath your elevation and the output from the Viewshed function is a useful technique for visualizing the relationship between visibility and terrain.

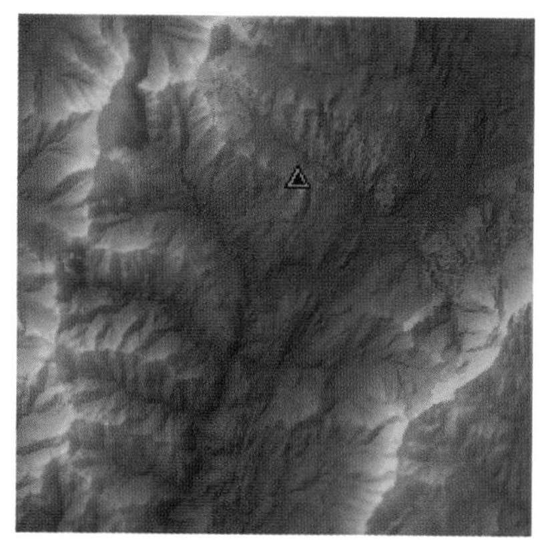

The relationship between visibility and the terrain

The Viewshed function can correct for the curvature of the earth and refraction of light rays passing through the atmosphere if the input surface has a projection file in which the ground units and surface z units are expressed in standard units—feet, meters, or units–meter.

Finding viewshed

The Viewshed function allows you to identify the places that can be seen from one or more observation points or lines. If lines are used as input, the observation points occur at the vertices of the lines.

The raster created from this function contains cells coded to indicate whether they are visible to or hidden from the observer. If there is more than one observer point, each visible cell in the raster shows the number of points from which it is visible.

Tip

Specifying a Z-factor

To get accurate viewshed results, the z units must be the same as the x,y units. If they are not the same, use a Z-factor to convert z units to x,y units. For example, if your x,y units are meters and your z units are feet, you could specify a Z-factor of 0.3048 to convert feet to meters.

Tip

Using optional parameters

Optional viewshed parameters—SPOT, OFFSETA, OFFSETB, and so on—will be used if they are present in the feature observer attribute table. For more information, search the ArcGIS Desktop Help system for Viewshed.

Creating a viewshed dataset

1. Click the Spatial Analyst dropdown arrow, point to Surface Analysis, and click Viewshed.
2. Click the Input surface dropdown arrow and click the input surface from which you want to calculate the viewshed.
3. Click the Observer points dropdown arrow and click the feature layer to use as observer points.
4. Optionally, check use Earth curvature.
5. Optionally, change the default Z-factor.
6. Optionally, change the default Output cell size.
7. Specify a name for the Output raster or leave the default to create a temporary dataset in your working directory.
8. Click OK.

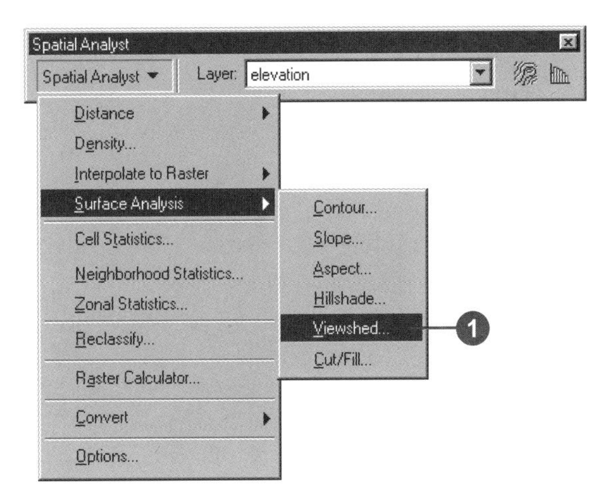

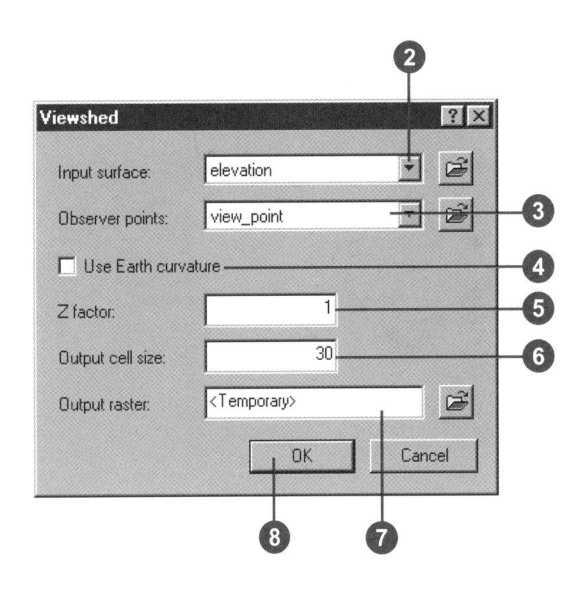

Cut/Fill

What is Cut/Fill?

Cut/Fill summarizes the areas and volumes of change between two surfaces. It identifies the areas and volume of the surface that have been modified by the addition or removal of surface material.

By taking two surfaces of a given area from two different time periods, the Cut/Fill function will produce a raster displaying regions of surface material addition, surface material removal, and areas where the surface has not changed over the time period. Negative volume values indicate areas that have been filled; positive volume values indicate regions that have been cut.

Taking river morphology as an example, to track the amount and location of erosion and deposition in a river valley, a series of cross sections need to be taken through the valley and surveyed on a regular basis to identify regions of sediment erosion and deposition.

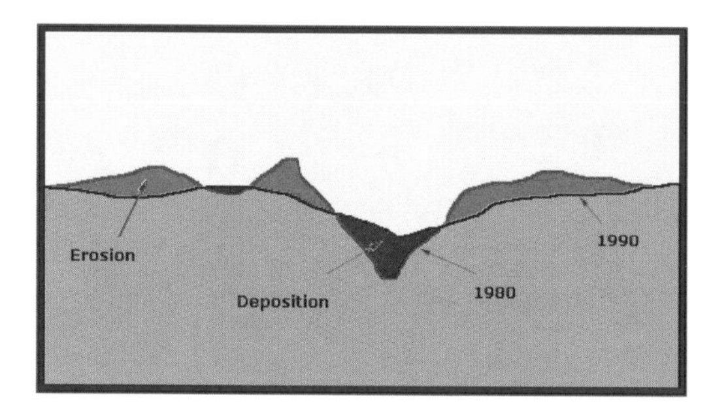

The Cut/Fill function peforms the surveying for you, identifying the areas that have been eroded, the areas of deposition, and the areas of no change. It also calculates the volume of surface material that has been cut or filled in each area.

The diagrams below show how the Cut/Fill function uses the Before and After surface to calculate the areas filled, cut, and areas that did not change as a result of the St Helen's volcano in the Pinochet Natioal Forest. Filled areas are displayed in green, cut areas in red, and areas where the surface material did not change are displayed in yellow on the Cut/Fill diagram.

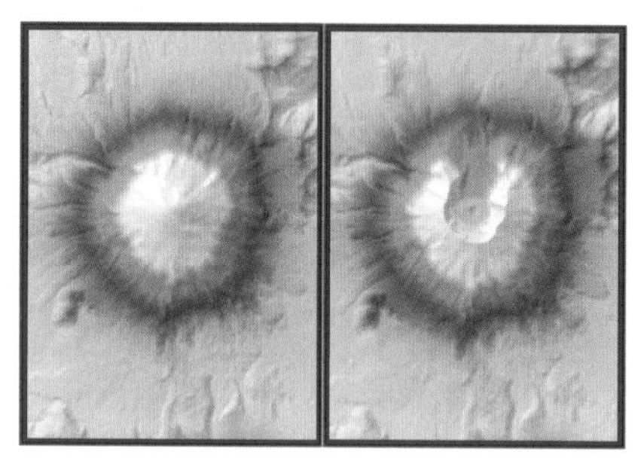

Before surface *After surface* *Cut/Fill surface*

Why use the Cut/Fill function?

With the Cut/Fill function you can:

- Identify regions of sediment erosion and deposition in a river valley.
- Calculate the volumes and areas of surface material to be removed and areas to be filled to level a site for building construction.
- Identify areas that become frequently inundated with surface material during a mudslide in a study to locate safe areas of stable land for building homes.

Calculating Cut/Fill

The Cut/Fill function enables you to create a map based on two input surfaces—before and after—displaying the areas and volumes of surface material that have been modified by the addition or removal of surface material.

The Z-factor is the number of ground x,y units in one surface z unit. The Input surface values are multiplied by the specified Z-factor to adjust the Input surface z units to another unit of measure.

Tip

Specifying a Z-factor

To get accurate Cut/Fill results, the z units must be the same as the x,y units. If they are not the same, use a Z-factor to convert z units to x,y units. For example, if your x,y units are meters and your z units are feet, you could specify a Z-factor of 0.3048 to convert feet to meters.

Creating a Cut/Fill dataset

1. Click the Spatial Analyst dropdown arrow, point to Surface Analysis, and click Cut/Fill.
2. Click the Before surface dropdown arrow and click a surface.
3. Click the After surface dropdown arrow and click another surface.
4. Optionally, change the default Z-factor.
5. Optionally, change the default Output cell size.
6. Specify a name for the Output raster or leave the default to create a temporary dataset in your working directory.
7. Click OK.

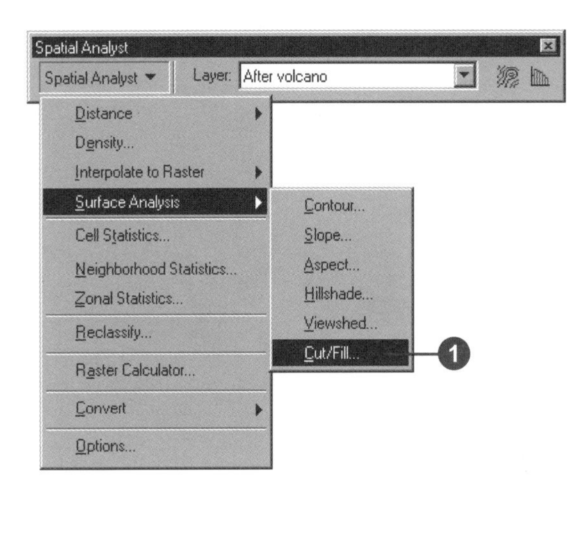

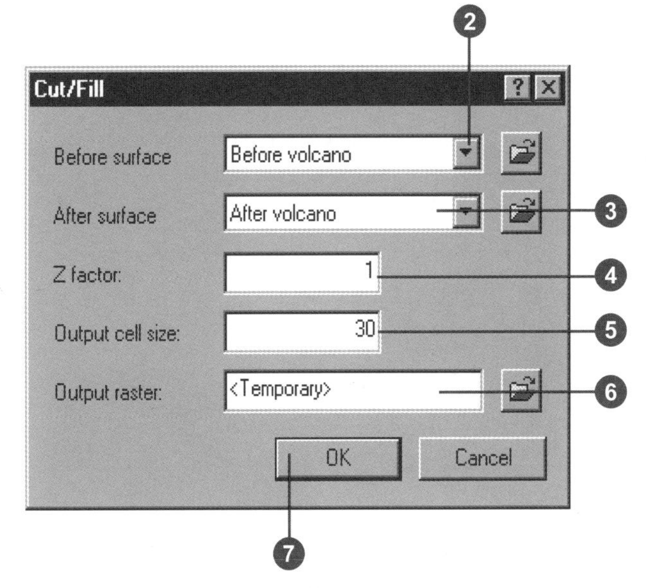

Cell statistics

What is the Cell Statistics function?

The *Cell Statistics* function is a local function, where the value at each location on the output raster is a function of the input values at each location.

By computing cell statistics, you can compute a statistic for each cell in an output raster that is based on the values of each cell of multiple input rasters.

Why calculate cell statistics?

Calculate cell statistics when you want to calculate a statistic between multiple rasters—for instance, to analyze a certain phenomenon over time, such as the average crop yield over a 10-year period or the range of temperatures between years.

In the graphic below, the variety of landuse types between cells of rasters from different years is calculated to identify the areas where the variety is greater than one—areas shaded gray. This indicates the areas that have changed landuse type over the time period, in this case, highlighting areas of urban sprawl—areas shaded red.

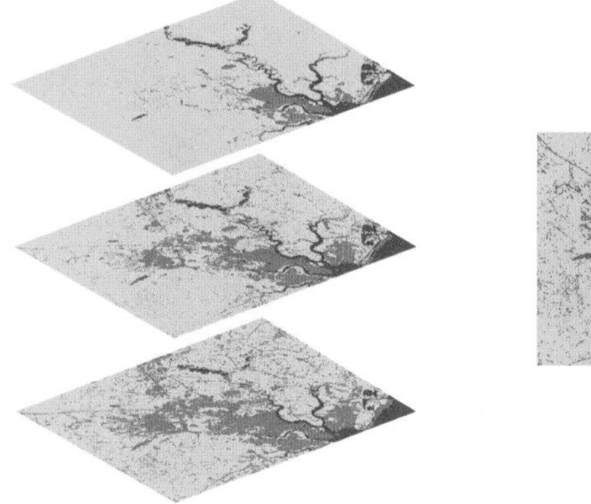

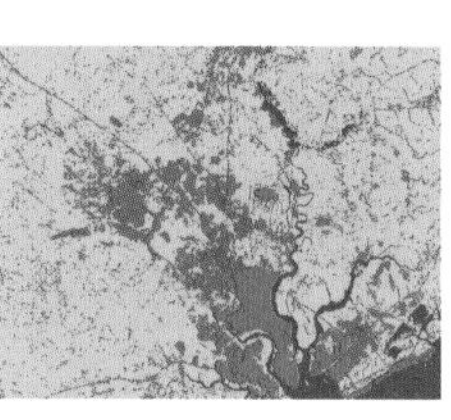

The following statistics can be computed on a cell-by-cell basis between input rasters, then sent to the corresponding cell location on the output raster:

Majority: determines the value that occurs most often on a cell-by-cell basis between inputs

Maximum: determines the maximum value on a cell-by-cell basis between inputs

Mean: computes the mean of the values on a cell-by-cell basis between inputs

Median: computes the median of the values on a cell-by-cell basis between inputs

Minimum: determines the minimum value on a cell-by-cell basis between inputs

Minority: determines the value that occurs least often on a cell-by-cell basis between inputs

Range: determines the range of values on a cell-by-cell basis between inputs

Standard Deviation: computes the standard deviation of the values on a cell-by-cell basis between inputs

Sum: computes the sum of the values on a cell-by-cell basis between inputs

Variety: determines the number of unique values on a cell-by-cell basis between inputs

Calculating cell statistics

The Cell Statistics function allows you to compute a statistic for each cell in an output raster based on the values of multiple input rasters.

If any cell in one of the input rasters contains a value of NoData, the value of the cell in the same location in the output raster will be NoData.

Tip

Setting analysis options

Click Options on the Spatial Analyst toolbar to set up your working directory, extent, and cell size for your analysis results.

Tip

Browsing for files or directories

If the file you need is not in your table of contents, or if you need to check the directory to place your output raster, click the Browse button.

Creating a dataset using Cell Statistics

1. Click the Spatial Analyst dropdown arrow and click Cell Statistics.
2. Click the Layers you want to use in the calculation; use the Shift key to highlight multiple layers.

   Alternatively, click the Browse button to access raster datasets on disk.
3. Click Add.
4. Click the Overlay statistic dropdown arrow and click the type of statistic you want to compute on your input layers.
5. Specify a name for the Output raster or leave the default to create a temporary dataset in your working directory.
6. Click OK.

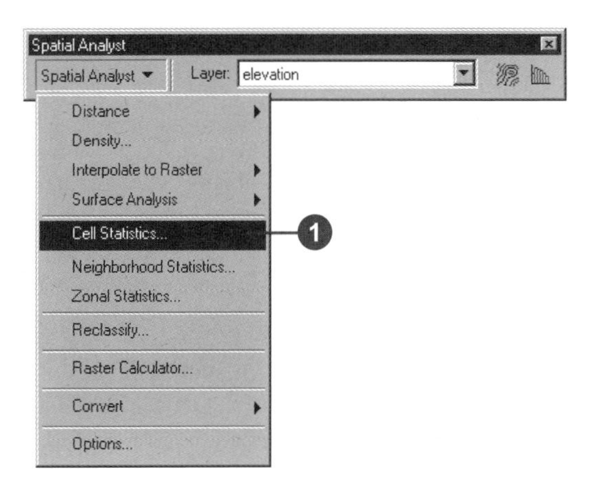

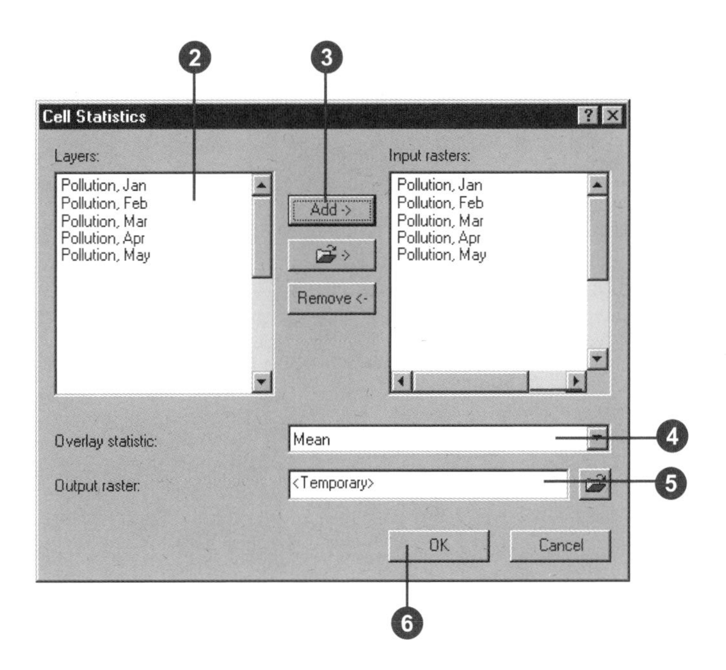

Neighborhood statistics

What is the Neighborhood Statistics function?

The *Neighborhood Statistics* function is a focal function that computes an output raster where the value at each location is a function of the input cells in some specified neighborhood of the location.

For each cell in the input raster, the Neighborhood Statistics function computes a statistic based on the value of the processing cell and the value of the cells within a specified neighborhood, then sends this value to the corresponding cell location on the output raster.

The following statistics can be computed within the neighborhood of each processing cell, then sent to the corresponding cell location on the output raster:

Majority: determines the value that occurs most often in the neighborhood

Maximum: determines the maximum value in the neighborhood

Mean: computes the mean of the values in the neighborhood

Median: computes the median of the values in the neighborhood

Minimum: determines the minimum value in the neighborhood

Minority: determines the value that occurs least often in the neighborhood

Range: determines the range of values in the neighborhood

Standard Deviation: computes the standard deviation of the values in the neighborhood

Sum: computes the sum of the values in the neighborhood

Variety: determines the number of unique values within the neighborhood

Neighborhood shapes

The neighborhoods that can be specified are a rectangle of any dimension, a circle of any radius, an annulus—a doughnut shape—of any radius, and a wedge in any direction.

Rectangle

The width and height units of a rectangular neighborhood can be in cells or in map units. The default is a neighborhood of 3-x-3 cells.

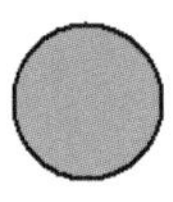

Circle

The size of the circle depends on the specified radius. The radius is identified in cells or map units, measured perpendicular to the x- or y-axis. Any cell center encompassed by the circle will be included in the processing of the neighborhood.

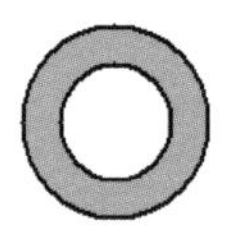

Annulus

Cells that fall within the annulus will be included in the processing of the neighborhood. The inner radius specifies the radius of the inner circle of the annulus from the center of the processing cell. Any cell falling within the radius will not be included in the processing of the neighborhood. The outer radius specifies the radius of the outer circle of the annulus from the center of the processing cell. The outer circle defines the extent of the neighborhood. Any cell center falling within the radius of the outer circle but outside the radius of the inner circle will be included in the processing of the neighborhood.

The radius is identified in cells or map units, measured perpendicular to the x- or y-axis.

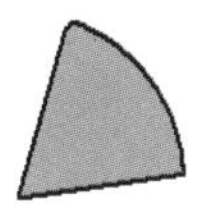

Wedge

Cells that fall within the wedge will be included in the processing of the neighborhood. The wedge is created by specifying a radius and an angle.

The radius is specified in either cell or map units, from the center of the processing cell, measured perpendicular to the x- or y-axis.

The start angle for the wedge can be an integer or floating point value from 0 to 360. Values for the wedge begin at 0 on the positive x-axis and increase counterclockwise until they return full-circle to 0.

The end angle for the wedge can be an integer or floating point value from 0 to 360. The angle defined by the start and end values is used to create the wedge. All cells that fall within the wedge are included in the processing of the neighborhood.

The neighborhood function for an individual cell

Take the processing cell with a value of 5 in the diagram that follows. With a rectangular 3-x-3 cell neighborhood, the sum of the value of neighboring cells plus the value of the processing cell equals 24. So a value of 24 is given to the value of the cell in the output raster in the same location as the processing cell in the input raster.

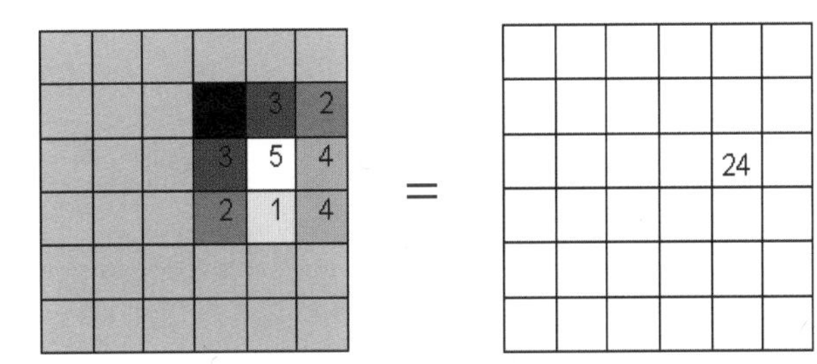

Input processing cells *Output value for one cell*

The neighborhood function on an entire dataset

Each cell in the output raster below has been calculated by summing the cells in a 3-x-3 neighborhood for each cell. The cells highlighted in yellow identify the neighborhood of the input processing cell with a value of 5 and output cell value of 24. This process is performed on every input processing cell to calculate an output value for each cell.

4	0	1	2	3	0
2	5	0		3	2
1	1	2	3	5	4
1	5	3	2	1	4
5		1	3	3	0
1	1	2	3	4	3

=

11	12	8	9	10	8
13	16	14	19	22	17
15	20	21	19	24	19
13	19	20	23	25	17
13	19	20	22	23	15
7	10	10	16	16	10

Input processing raster *Output raster*

Processing cells of NoData

If a cell of NoData is present in the neighborhood, it will be ignored in the processing. If the entire neighborhood consists of cells of NoData, the output cell value will be NoData.

Why calculate neighborhood statistics?

Calculating neighborhood statistics is useful for obtaining a value for each cell based on a specified neighborhood. For example, when examining ecosystem stability, it might be useful to obtain the variety of species for each neighborhood to identify the locations that are lacking a variety of species.

The example below takes a raster displaying land cover and calculates the variety of different land cover types in each neighborhood.

Land cover type

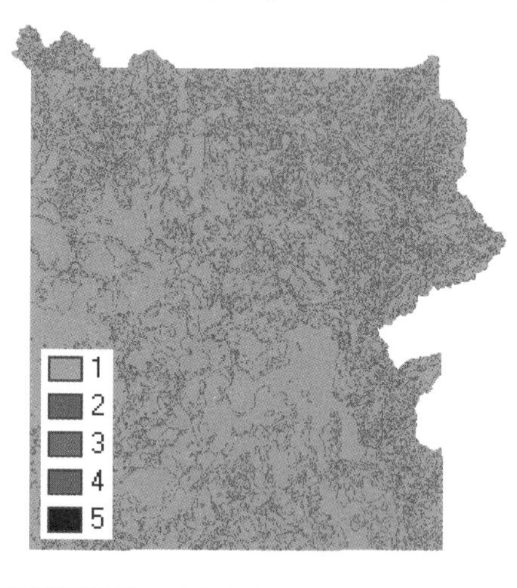

A value is given to each cell in the output raster based on the cell's value and the values in a specified neighborhood. A quick overview of the region shows you the areas with more than one land cover.

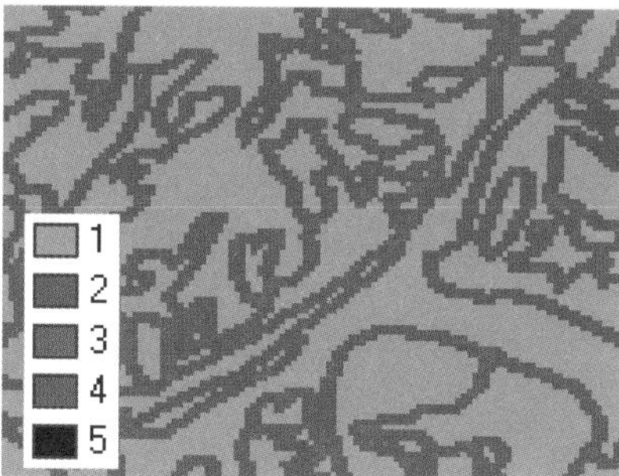

By zooming in on a particular region, you can see more clearly that there are neighborhoods with three and even four types of land cover.

Attributes of NbrVariety of land cover

ObjectID*	Value	Count
1	1	2771514
2	2	758855
3	3	24495
4	4	169
5	5	1

Record: 1 Show: All Selected

The attribute table for the output raster tells you how many cells there are that contain multiple types of land cover within the specified neighborhood.

Calculating neighborhood statistics

The Neighborhood Statistics function allows you to calculate a statistic for each cell based on the value of that cell and the values in a neighborhood you specify. Use it, for example, to find the most dominant species in a neighborhood—majority—or to see how many species are located in each neighborhood—variety.

The Neighborhood dictates the shape of the area used to obtain a value for the cell being processed. The Neighborhood Settings dictate the size of the shape—the number of cells or map units—that is used to obtain a value for the cell being processed.

Tip

Highlighting cells on the map

Right-click the output raster and click Open Attribute Table. Click a row in the table to highlight the cells on the map.

Creating a map using neighborhood statistics

1. Click the Spatial Analyst dropdown arrow and click Neighborhood Statistics.
2. Click the Input data dropdown arrow and click the layer on which you want to perform neighborhood statistics.
3. Click the Field dropdown arrow and click the field from the Input data you wish to use.
4. Click the Statistic type dropdown arrow and click the type of statistic you wish to compute.
5. Click the Neighborhood dropdown arrow and click the type of neighborhood you wish to use.
6. Specify the Neighborhood Settings for your chosen neighborhood.
7. Optionally, change the default Output cell size.
8. Specify a name for the Output raster or leave the default to create a temporary dataset in your working directory.
9. Click OK.

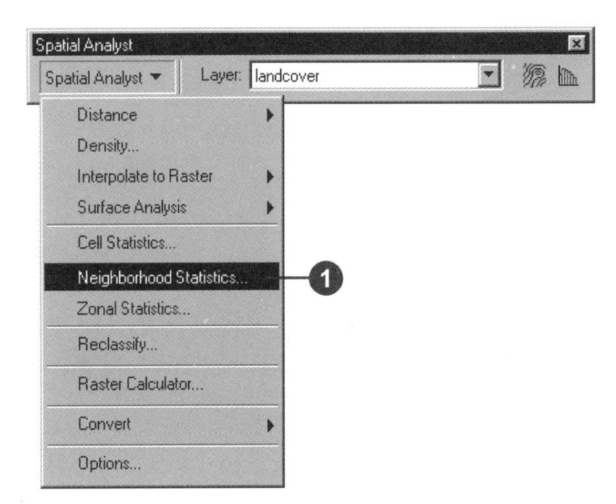

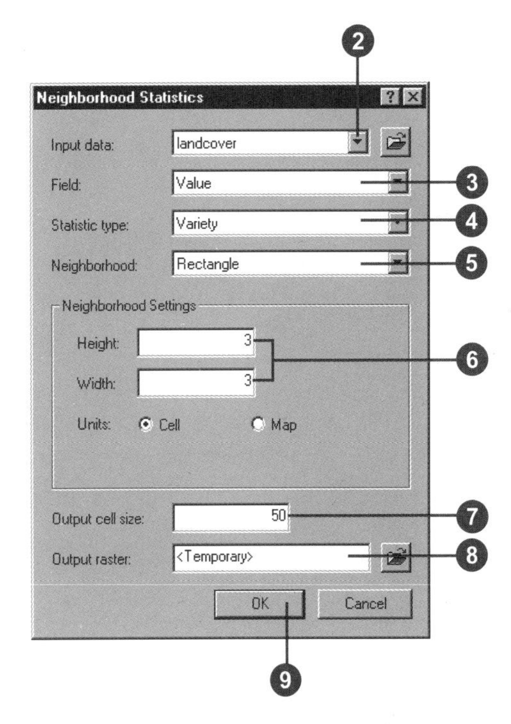

Zonal statistics

What is the Zonal Statistics function?

With the Zonal Statistics function, a statistic is calculated for each zone of a zone dataset, based on values from another dataset.

A zone is all the cells in a raster that have the same value, regardless of whether or not they are contiguous. However, both raster and feature datasets can be used as the zone dataset. So, for example, residential is a zone of a landuse raster dataset, or a roads feature dataset can be the zone for an accident dataset.

Zonal statistical functions perform operations on a per-zone basis; a single output value is computed for every zone in the input zone dataset.

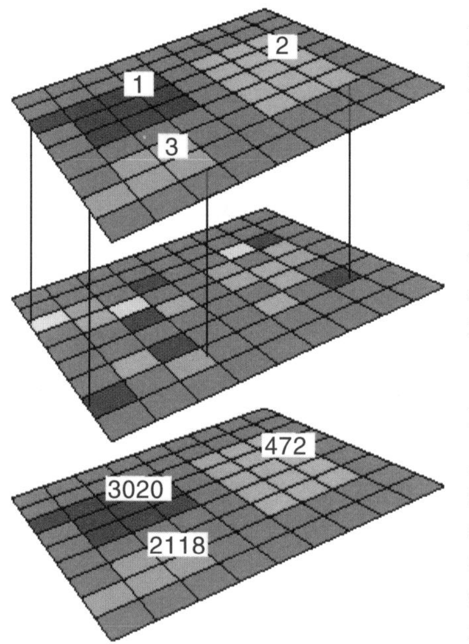

Zone layer:

Defines the zones—shape, values, and locations.

Value raster:
Contains the input values used in calculating the output for each zone.

Input Zone layer:

A field can be added to the zone layer attribute table, containing the statistics calculated for each zone.

The following statistics can be computed within each zone:

Majority: determines the value that occurs most often in the zone

Maximum: determines the maximum value in the zone

Mean: computes the mean of the values in the zone

Median: computes the median of the values in the zone

Minimum: determines the minimum value in the zone

Minority: determines the value that occurs least often in the zone

Range: determines the range of values in the zone

Standard Deviation: computes the standard deviation of the values in the zone

Sum: computes the sum of the values in the zone

Variety: determines the number of different values within the zone

Why use zonal statistics?

You might calculate the mean elevation for each forest zone or the number of accidents along each of the roads in a town. Alternatively, you might want to know how many different types of vegetation there are in each elevation zone—variety. The graphics below show an example of the inputs and outputs from the Zonal Statistics function. The variety of vegetation species per elevation zone is displayed in the output table and chart. The most variety of species occurs at elevation levels of around 2,500 meters.

Input zone dataset: elevation zones (Elevation range from 1,547 to 3,358 meters)

Attributes of Zones

VALUE	VARIETY
1547	13
1773	28
1999	41
2226	47
2452	50
2679	43
2905	26
3132	14
3358	3

Record: 0

Output table

Input value raster: vegetation type

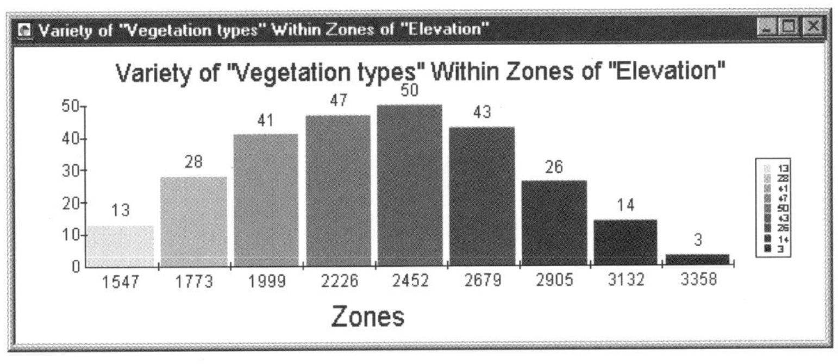

Output chart

Calculating zonal statistics

The Zonal Statistics function allows you to compute statistics for each zone of a zone dataset based on the information in a value raster. This could be average population density per zone of pollution or vegetation type per zone of elevation.

The zone dataset can be feature or raster data. The value raster must be a raster dataset.

Tip

Using NoData

Uncheck the Ignore NoData in calculations check box if you want NoData values to be included in the calculation. If there are NoData values within a zone, the output value for the zone will be NoData because there is insufficient information to complete the calculation.

Leave the Ignore NoData in calculations check box checked if you want NoData values to be ignored. Only cells on the value raster that have data values within each zone will be used in the calculation.

Creating a chart using zonal statistics

1. Click the Spatial Analyst dropdown arrow and click Zonal Statistics.
2. Click the Zone dataset dropdown arrow and click the layer you want to use.
3. Click the Zone field dropdown arrow and click the field of the Zone layer you wish to use.
4. Click the Value raster dropdown arrow and click the raster you wish to use.
5. Uncheck Ignore NoData in calculations to use the NoData values of the Value raster in the calculation.
6. Check the check box to Join the output table to the zone layer.

   Note that this option is only available for layers, not datasets you browsed to.
7. Click the Chart statistic dropdown arrow and click the type of statistic you wish to chart.
8. Specify a name for the Output table or leave the default to create a table in your working directory.
9. Click OK.

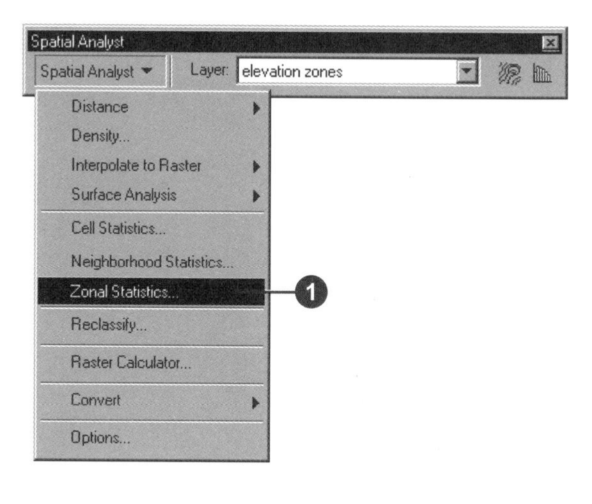

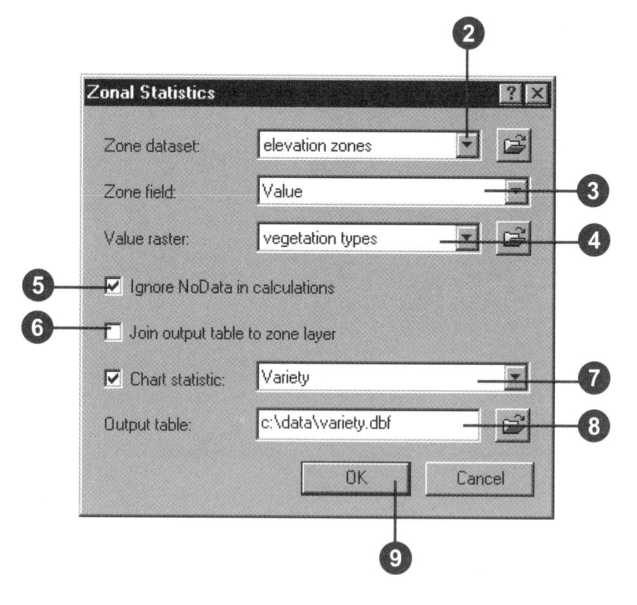

Reclassification

What is reclassification?

Reclassifying your data simply means replacing input cell values with new output cell values.

The input data can be any supported raster format. If you add a multiband raster, the first band will be taken and used in the reclassification.

Why reclassify your data?

There are many reasons why you might want to reclassify your data. Some of the most common reasons are:

- To replace values based on new information
- To group certain values together
- To reclassify values to a common scale—for example, for use in suitability analysis or for creating a cost raster for use in the Cost Weighted Distance function
- To set specific values to NoData or to set NoData cells to a value

Replacing values based on new information

Reclassification is useful when you want to replace the values in the input raster with new values. This could be due to finding out that the value of a cell or a number of cells should actually be a different value, for example, the landuse in an area changed over time.

Grouping values together

You may want to simplify the information in a raster. For instance, you may want to group together various types of forest into one forest class.

Reclassifying values of a set of rasters to a common scale

Another reason to reclassify is to assign values of preference, sensitivity, priority, or some similar criteria to a raster. This may be done on a single raster—a raster of soil type may be assigned values of 1–10 that represent erosion potential—or with several rasters to create a common scale of values.

For example, when finding slopes most at risk of avalanche activity, input rasters might be slope, soil type, and vegetation. Each of these rasters might be reclassified on a scale of 1–10 depending on the susceptibility of each attribute in each raster to avalanche activity—that is, steep slopes in the slope raster might be given a value of 10 because they are most susceptible to avalanche activity. For more details on suitability modeling, see 'Finding a site for a new school in Stowe, Vermont, USA', in Chapter 2.

Setting specific values to NoData or setting NoData cells to a value

Sometimes you want to remove specific values from your analysis. This might be, for example, because a certain landuse type has restrictions, such as wetland restrictions, which means you cannot build there. In such cases, you might want to change these values to NoData in order to remove them from further analysis.

In other cases, you may want to change a value of NoData to be a value, such as in the case where new information means a value of NoData has become a known value.

Reclassifying your data

The Reclassify dialog box enables you to modify the values in an input raster and save the changes to a new output raster.

There are many reasons why you may wish to do this, including replacing values based on new information, grouping entries, reclassifying values to a common scale—for example, for use in suitability analysis—or setting specific values to NoData or setting NoData cells to a value.

The Load button enables you to load a remap table that was previously created by pressing the Save button and apply it to the input raster. ▸

Tip

Replacing NoData values

NoData values can be turned into numeric values in the same way as replacing any value.

Tip

Changing the classes of your old values

Click Classify to classify your old values differently. Click Unique to separate classes of old values into unique values.

Replacing values based on new information

1. Click the Spatial Analyst dropdown arrow and click Reclassify.
2. Click the Input raster dropdown arrow and click the raster with values you wish to change.
3. Click the Reclass field dropdown arrow and click the field you wish to use.
4. Click the New values you wish to change and type a new value.
5. Click all the other New values while holding down the Shift key, then click Delete Entries.

   All other values will remain the same in the Output raster.
6. Optionally, click Save to save the remap table.
7. Specify a name for the Output raster or leave the default to create a temporary dataset in your working directory.
8. Click OK.

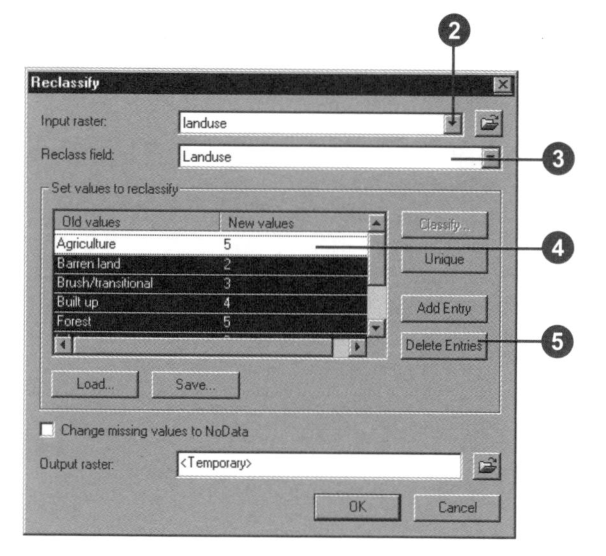

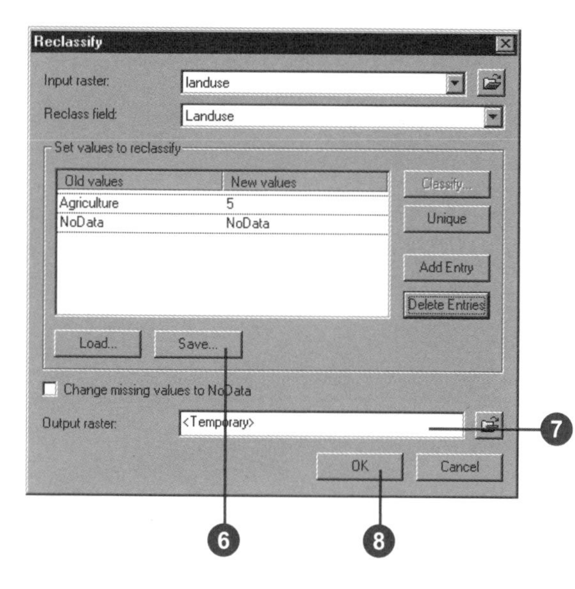

The Save button enables you to save a remap table for later use.

Tip

Ungrouping entries

To ungroup entries, right-click the group and click Ungroup Entries.

Tip

Changing the classes of your old values

Click Classify to change the classification of your old values. Click Unique to separate classes of old values into unique values.

Grouping entries

1. Click the Spatial Analyst dropdown arrow and click Reclassify.
2. Click the Input raster dropdown arrow and click the raster with values you wish to group.
3. Click the Reclass field dropdown arrow and click the field you wish to use.
4. Click the Old values you wish to group—click one, then hold down the Shift key and click the next one—then right-click and click Group Entries.
5. Give the grouped entry and other Old values the New values you wish for them to have.
6. Optionally, click Save to save the remap table.
7. Specify a name for the Output raster or leave the default to create a temporary dataset in your working directory.
8. Click OK.

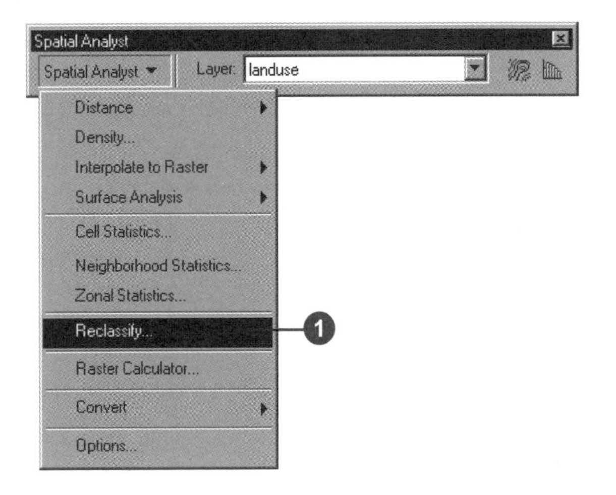

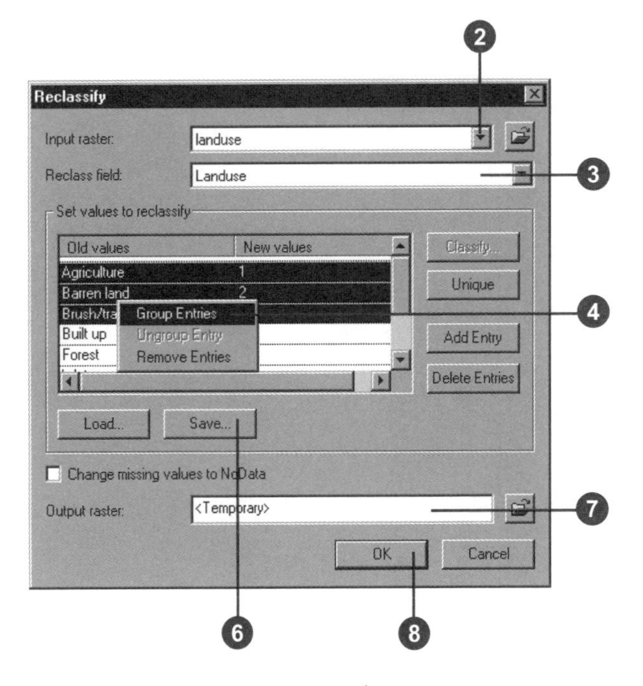

Use the Reclassify dialog box to apply a common scale of values to your rasters when doing suitability modeling. This involves reclassifying each raster on the same scale, giving higher new values to those old values that are more important to consider.

Tip

Changing the classes of your old values

Click Classify to change the classification of your old values. Click Unique to separate classes of old values into unique values.

See Also

For more details on suitability modeling, see 'Finding a site for a new school in Stowe, Vermont, USA', in Chapter 2.

Reclassifying values of a set of rasters to a common scale

1. Click the Spatial Analyst dropdown arrow and click Reclassify.
2. Click the Input raster dropdown arrow and click the raster with values you wish to prioritize.
3. Click the Reclass field dropdown arrow and click the field you wish to use.
4. Click the New values input box for each entry and prioritize the entries; this is subjective according to your spatial problem—for example, preference, cost, or time.
5. Optionally, click Save to save the remap table.
6. Specify a name for the Output raster or leave the default to create a temporary dataset in your working directory.
7. Click OK.

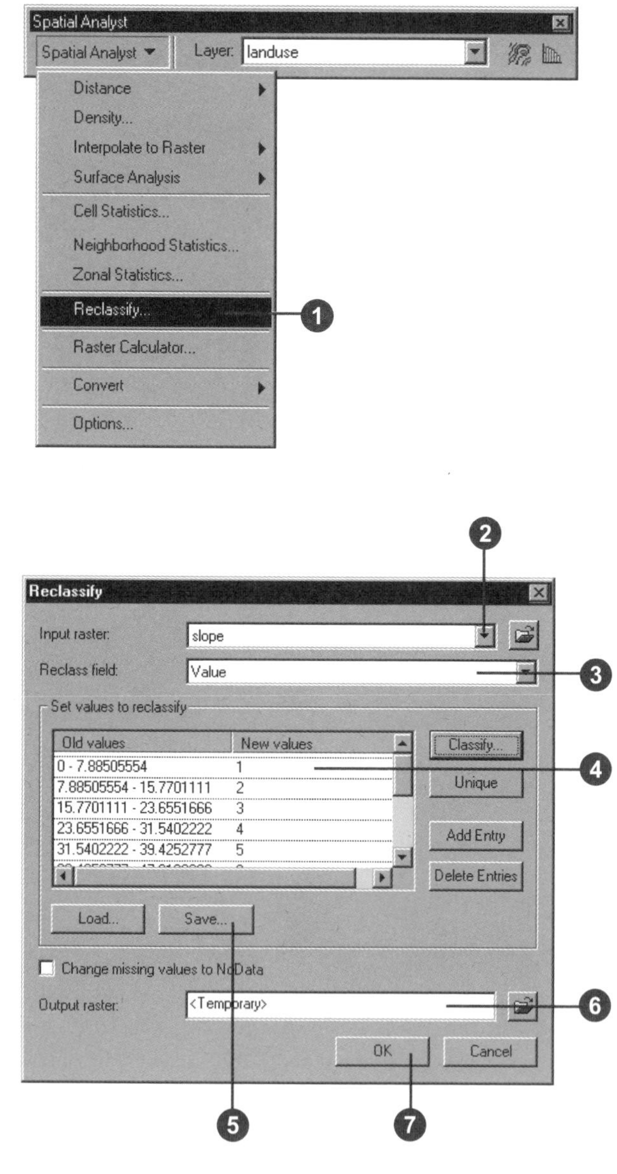

Tip

Changing input ranges to be unique values

If your input values are split into ranges and you want them to be unique values, click Unique.

See Also

See 'Standard classification schemes' in Using ArcMap *for information on classification schemes.*

Changing the classification of input ranges

1. Click the Spatial Analyst dropdown arrow and click Reclassify.
2. Click the Input raster dropdown arrow and click the raster with values you wish to reclassify.
3. Click the Reclass field dropdown arrow and click the field you wish to use.
4. Click the Classify button.
5. Click the Method dropdown arrow and choose a classification method to use to reclassify your input data.
6. Click the Classes dropdown arrow and choose the number of classes into which your input data will be split.
7. Click OK.
8. Modify the New values for your Output raster if appropriate.
9. Specify a name for the Output raster or leave the default to create a temporary dataset in your working directory.
10. Click OK on the Reclassify dialog box.

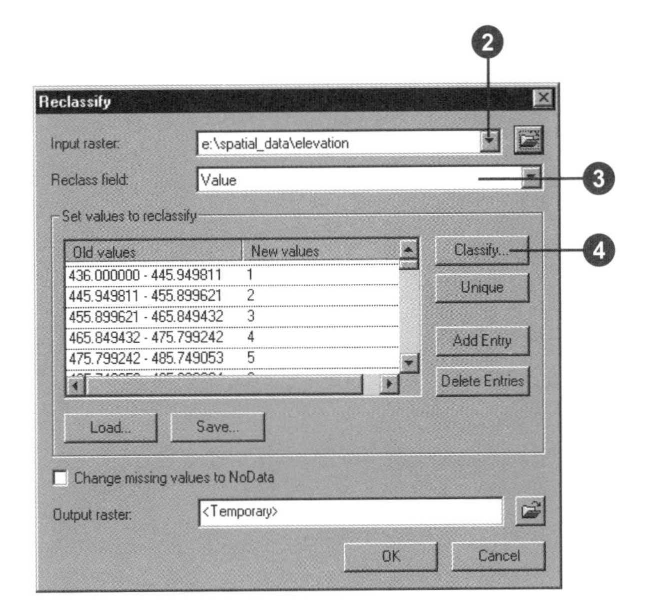

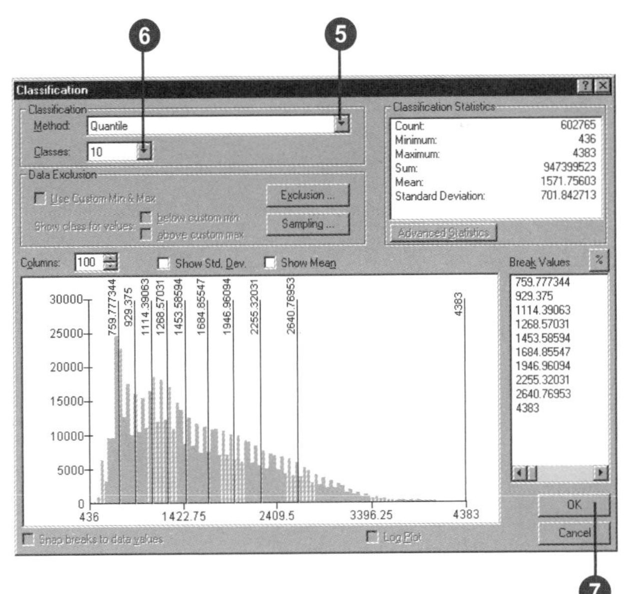

Tip

Changing a value to NoData

You can type "NoData" in the input box for a new value to change an input value to NoData.

Setting specific values to NoData

1. Click the Spatial Analyst dropdown arrow and click Reclassify.
2. Click the Input raster dropdown arrow and click the raster with values you wish to set to NoData.
3. Click the Reclass field dropdown arrow and click the field you wish to use.
4. Click the input boxes for the New values you wish to change to NoData.
5. Click Delete Entries.
6. Check Change missing values to NoData.
7. Optionally, click Save to save the remap table.
8. Specify a name for the Output raster or leave the default to create a temporary dataset in your working directory.
9. Click OK.

   The values you deleted will be changed to NoData in the Output raster.

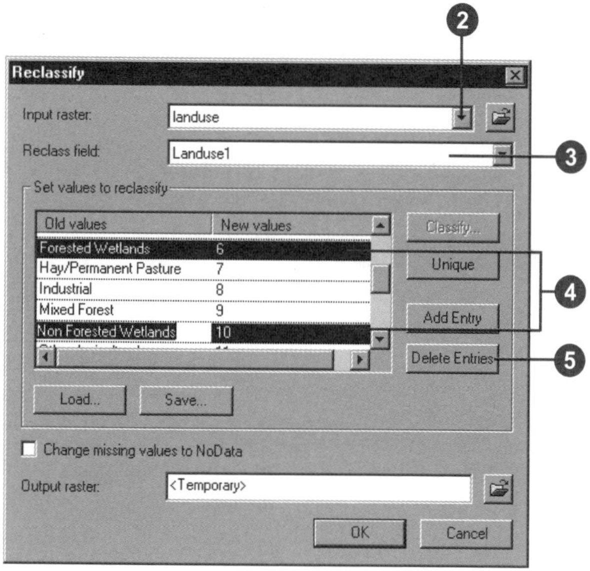

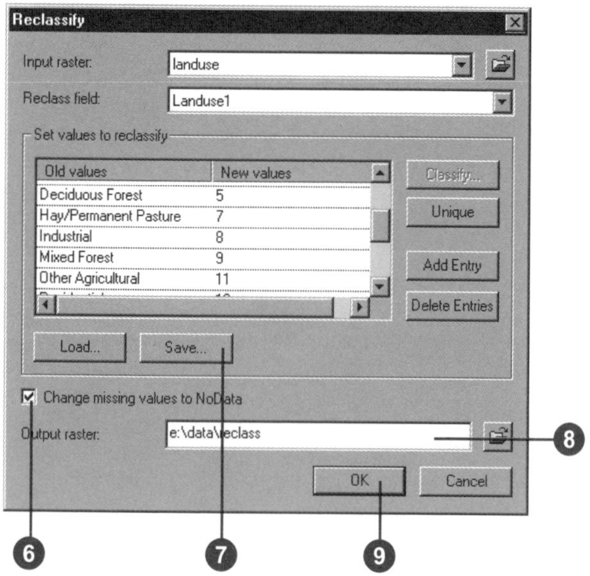

The Raster Calculator

What can you do with the Raster Calculator?

The Raster Calculator provides you with a powerful tool for performing multiple tasks. You can type in *Map Algebra* syntax to perform mathematical calculations using *operators* and *functions*, set up selection queries, or type in Spatial Analyst function syntax. Inputs can be grid datasets or raster layers, shapefiles, coverages, tables, constants, and numbers.

Mathematical operators and functions

Operators and functions evalulate the expression only for input cells that are spatially coincident with the output cell.

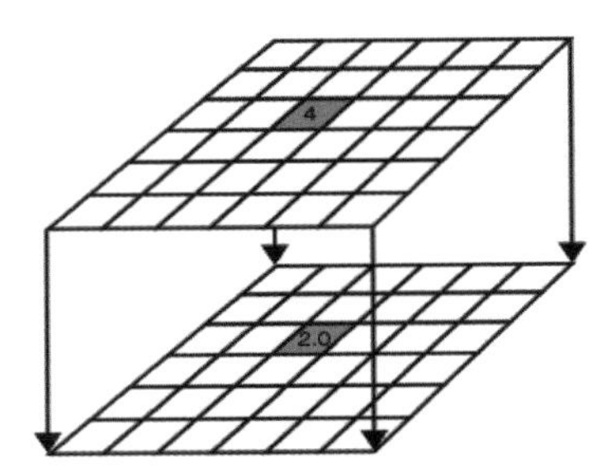

Sqrt([Inlayer1])

Mathematical operators

Mathematical operators apply a mathematical operation to the values in two or more input rasters. Three groups of mathematical operators are available in the Raster Calculator: *Arithmetic*, *Boolean*, and *Relational*. All operators, including Bitwise, Combinatorial, and Logical, can be typed into the Raster Calculator. For supported operators and precedence values, see Appendix B.

Arithmetic operators

Arithmetic operators allow for the addition, subtraction, multiplication, and division of two rasters or numbers, or a combination of the two.

Arithmetic operators: *, /, -, +

For example, the result of [Inlayer1] + [Inlayer2] / 2 results in an output raster displaying the mean value for every cell.

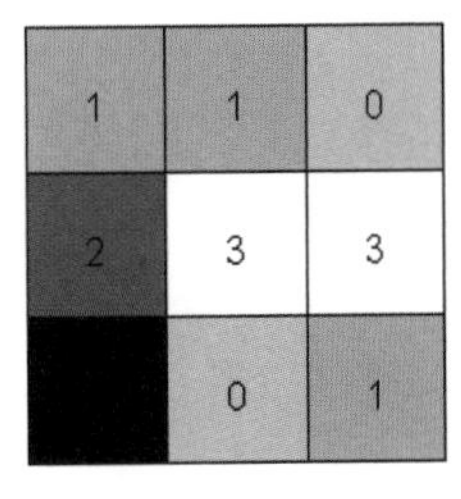

Input raster Inlayer1

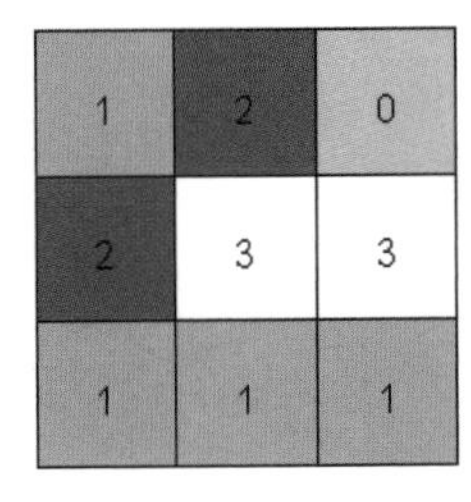

Input raster Inlayer2

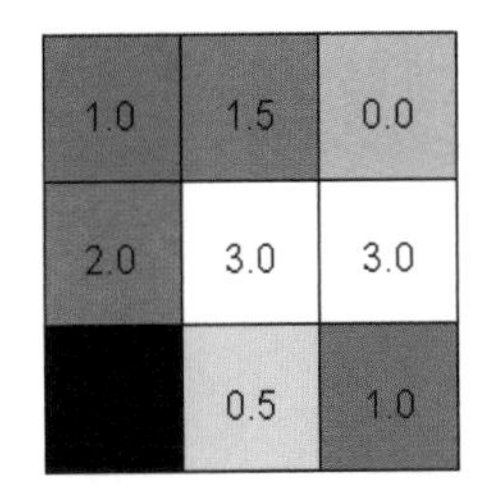

Output raster—mean of Inlayer1 and Inlayer2

Boolean operators

Boolean operators use Boolean logic—TRUE or FALSE—on input rasters on a cell-by-cell basis. Output values of TRUE are written as 1 and FALSE as 0.

Boolean operators: And, Or, Xor, Not

And (&): Finds where values are true (nonzero) in the cells of both input rasters.

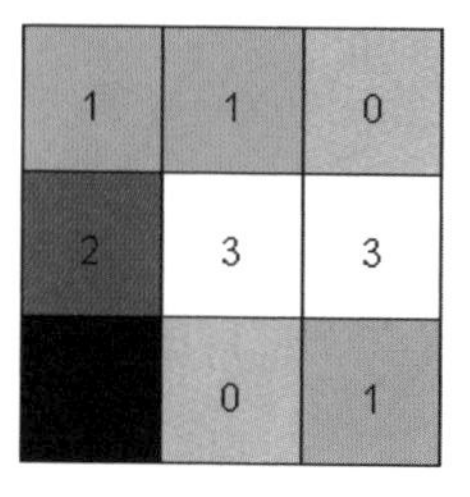

Input raster Inlayer1

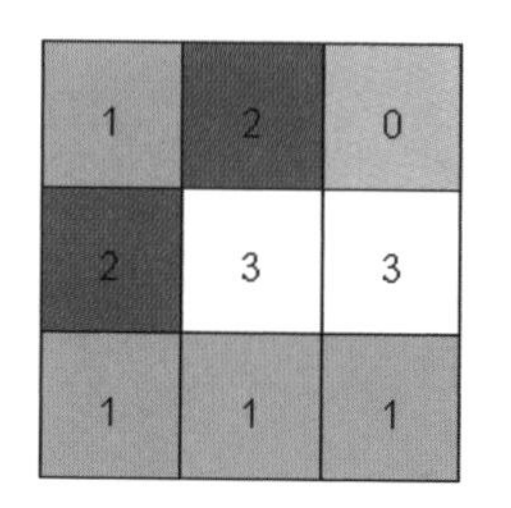

Input raster Inlayer2

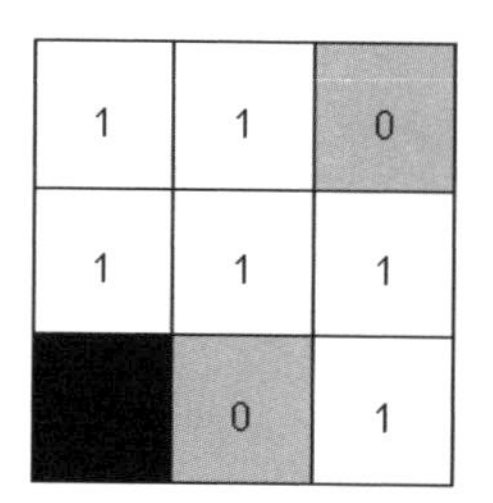

Output raster [Inlayer1] & [InLayer2]

Or (|): Finds where nonzero values are present in the cells of one or both input rasters.

Xor (!): Finds where nonzero values are present in the cells of one input raster or another input raster, but not both.

Not (^): Finds where nonzero values are not present in the cells of a single input raster.

Relational operators

Relational operators evaluate specific relational conditions. If a condition is TRUE, the output is assigned 1; if a condition is FALSE, the output is assigned 0.

Relational operators: ==, >, <, <>, >=, <=

For example, the result of Inlayer1 <> 3, where values in Inlayer1 are not equal to 3, might give an output raster showing all other landuses except forest, if forest was represented by a value of 3.

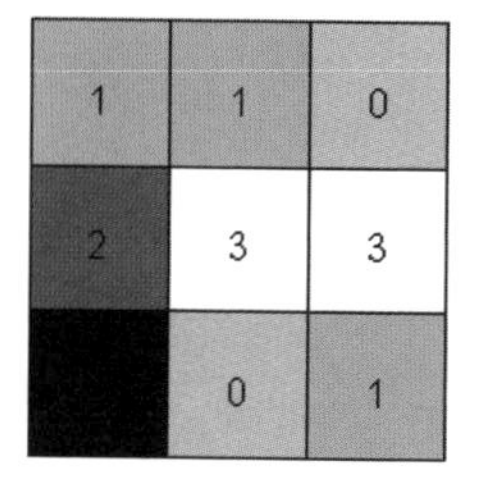

Input raster Inlayer1

1 1 1
1 0 0
1 1

Output raster [Inlayer1] <> 3

Mathematical functions

Mathematical functions are applied to the values in a single input raster.

There are four groups of mathematical functions: Logarithmic, Arithmetic, Trigonometric, and Powers.

Logarithmic functions: The *Logarithmic functions* perform exponential and logarithmic calculations on input rasters and numbers. The base e (Exp), base 10 (Exp10), and base 2 (Exp2) exponential capabilities and the natural (Log), base 10 (Log10), and base 2 (Log2) logarithmic capabilities are available.

For example, the result of Exp([Inlayer1]) is shown below:

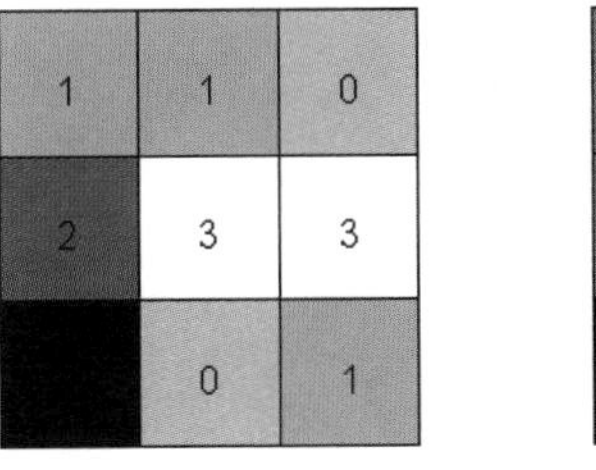

Input raster Inlayer1

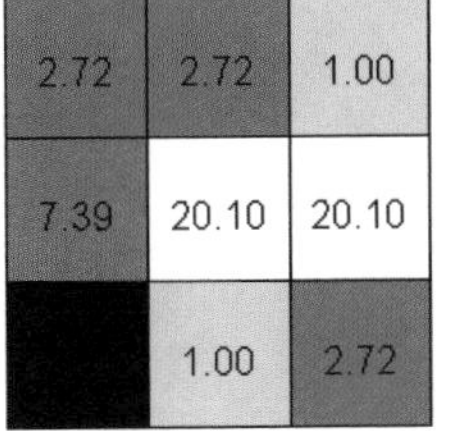

Output raster Exp([Inlayer1])

Arithmetic functions: There are six *Arithmetic functions*. The Abs function takes the absolute value of the values in an input raster. Two rounding functions, Ceil and Floor, convert decimal-point values into whole numbers. Int and Float convert values from and to integer and floating-point values. The IsNull function returns 1 if the values on the input raster are NoData and 0 if they are not.

Trigonometric functions: The *Trigonometric functions* perform various trigonometric calculations on the values in an input raster. The sine (Sin), cosine (Cos), tangent (Tan), inverse sine (Asin), inverse cosine (Acos), and inverse tangent (Atan) functions exist.

Power functions: Three *Power functions* are available. The square root (Sqrt) of the values on the input raster can be calculated, the square (Sqr) determined, or the values raised to a power (Pow).

Map Algebra syntax

Map Algebra is the analysis language for Spatial Analyst. An output will result from some manipulation of the input. Type Map Algebra syntax to access a variety of functions.

Basic rules and limitations

- Inputs can be grid datasets, raster layers, shapefiles coverages, tables, constants, or numbers.
- Outputs can be grid datasets, shapefiles, tables and files stored on disk, such as ASCII files.
- Multiline expressions are supported.
- The raster layer name, if the raster layer is in the table of contents, or the full pathname to the dataset must be provided.
- The accumulative operators are not supported.
- Spatial Analyst functions can be run in the Raster Calculator:

slope([Inlayer1])—calculates the slope of Inlayer1

mean([Inlayer1], [Inlayer2], [Inlayer3])—calculates the mean value between rasters on a cell-by-cell basis

hillshade(e:\spatial\ingrid)—creates a hillshade from a grid dataset on disk

For more on Map Algebra, see Appendix A, or search the ArcGIS Desktop Help index for Map Algebra.

Using the Raster Calculator

The Raster Calculator gives you access to numerous tools. Use Map Algebra to weight rasters and combine them as part of a suitability model, make selections on your data in the form of queries, apply mathematical operators and functions, or type Spatial Analyst functions.

Use layers from the table of contents, or type the full pathname to the grid dataset, shapefile, coverage, or table on disk. For example, typing c:\spatial\elevation * 2 will use the elevation dataset in the location specified and multiply it by two. ►

Tip

Changing the font used to build an expression

Right-click inside the expression box and click Font.

Tip

Creating permanent output

Either specify an output name in the expression,

*outgrid = [inlayer] * [inlayer2]*

or create a temporary result, then right-click the output in the table of contents and click Make Permanent.

Using the Raster Calculator to weight rasters

1. Click the Spatial Analyst dropdown arrow and click Raster Calculator.
2. Double-click the layer to which you want to add weight.

   The layer will be added to the expression box.
3. Click the Multiply button.
4. Type a value to weight the dataset.
5. Follow steps 1 through 5 for all datasets you want to weight.
6. Click Evaluate.

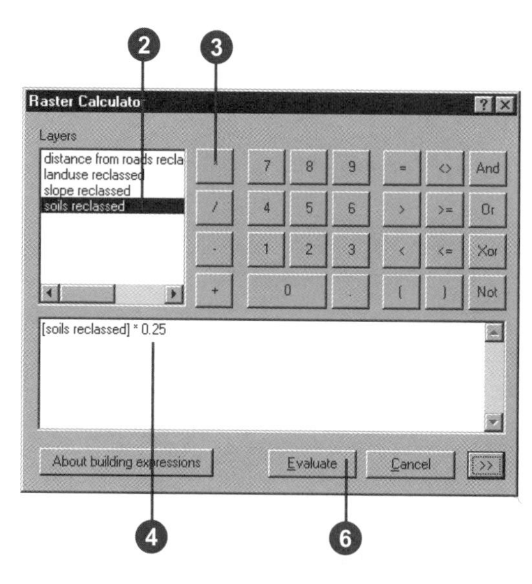

Using the Raster Calculator to combine rasters

1. Click the Spatial Analyst dropdown arrow and click Raster Calculator.
2. Double-click the first layer.
3. Click the Add button.
4. Click the next layer.
5. Repeat steps 3 and 4 to add all your datasets together.
6. Click Evaluate.

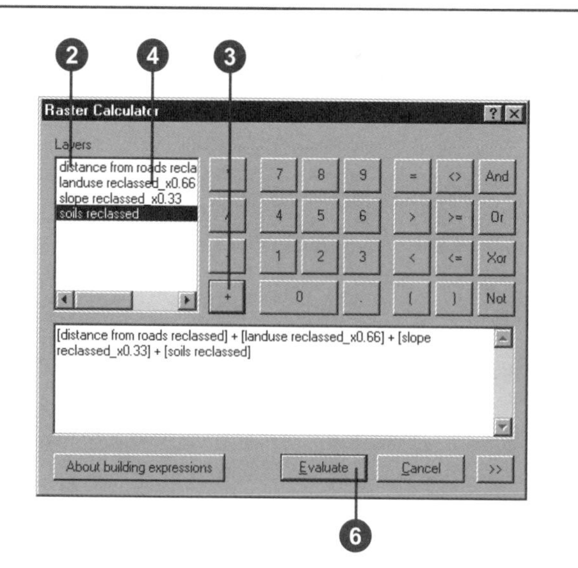

The Raster Calculator enables you to perform many different types of queries on your data. For example,

[elevation] > 3000 & [landuse] == 5

will identify all those cells of the elevation raster that are higher than 3,000 meters with a landuse value of 5. Cells that meet the criteria—cells over 3,000 meters with a landuse value of 5—will be given a value of 1 in the output raster. Cells that do not meet the criteria—cells with elevations lower than 3,000 meters and with a landuse value that is not 5—will be given a value of 0. ►

Tip

Accessing recently used expressions

Right-click inside the expression box and click Recent expressions. Copy/paste expressions into the expression box.

Tip

Writing long expressions

If your expression is too long to fit on a single line, use the line continuation symbol, ~, at the end of the line, then continue to type your expression on the next line.

See Also

Type expressions, multiline in the online Help index for information on entering multiline expressions.

Using the Raster Calculator to make selections on your data

1. Click the Spatial Analyst dropdown arrow and click Raster Calculator.
2. Double-click the layer you want to make a selection from to add it to the Expression box—for example, elevation.
3. Click the operator you wish to use—for example, ">" or "And".
4. Click or type a value—for example, 3000—or click another Layer, depending on the operator you choose.
5. Click Evaluate to perform the calculation.

   The Raster Calculator will be closed and the result displayed.

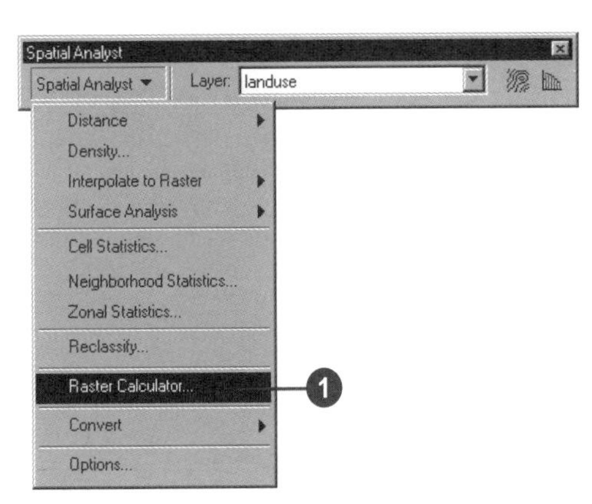

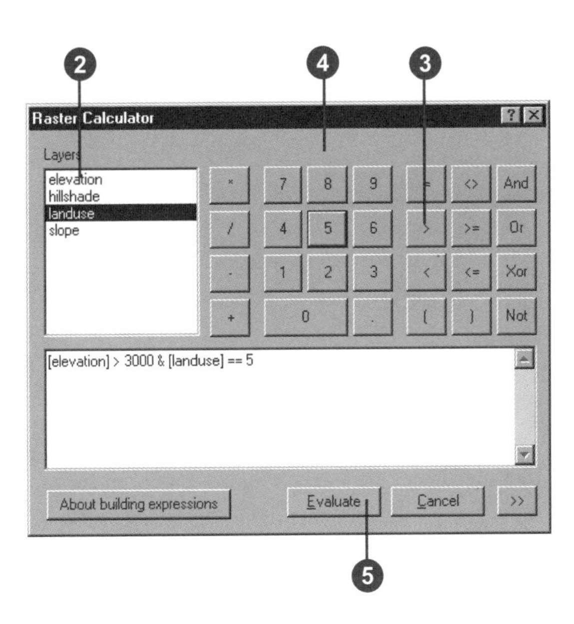

Use the Raster Calculator to perform mathematical functions on your data.

The example used in the task is Exp.

Use the following syntax when using raster layers from the Layers list (those layers you added to ArcMap):

Exp([Density])

If you have not added the raster dataset you wish to use as a raster layer to ArcMap, it will not appear as a layer in the Layers list. If the raster dataset is a grid, you can specify the path to the grid dataset on disk:

Exp(c:\data\Density) ▸

Tip

Expanding the expression box

Click and drag the bottom of the Raster Calculator dialog box to expand the expression box.

See Also

For more information on Map Algebra syntax and rules, see Appendix A. For supported operators and precedence values, see Appendix B.

See Also

Click About Building Expressions to access help for the Raster Calculator, including the Spatial Analyst functional reference.

Using the Raster Calculator to perform mathematical functions on your data

1. Click the Spatial Analyst dropdown arrow and click Raster Calculator.
2. Click the Expansion button to expand the Raster Calculator and reveal the mathematical functions.
3. Click the function you want to use.
4. Double-click the layer to which you want to apply the function.
5. Click Evaluate.

Spatial Analyst
Spatial Analyst ▾ Layer: landuse
Distance
Density...
Interpolate to Raster
Surface Analysis
Cell Statistics...
Neighborhood Statistics...
Zonal Statistics...
Reclassify...
Raster Calculator...
Convert
Options...
1

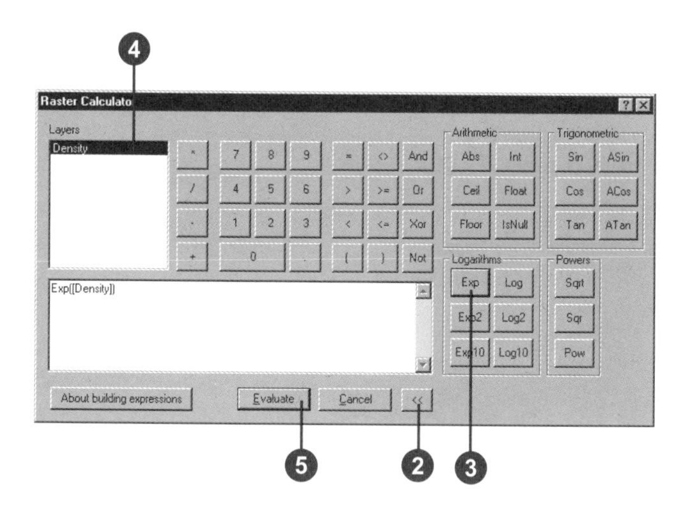

Use the Raster Calculator to perform Spatial Analyst functions. For example:

slice([elevation], eqinterval, 20)

SLICE splits the input data—elevation—into 20 equal interval classes.

For all Map Algebra expressions, specify the Map Algebra function, an open bracket, the input raster, and then any other parameters and a closed bracket.

Type the following syntax when using raster layers from the Layers list—those layers you added to ArcMap:

slice([elevation], eqinterval, 20)

Otherwise, if the input data is a coverage, shapefile, table, or grid dataset, you can specify the path to the input data on disk:

slice(c:\data\elevation], eqinterval, 20)

Tip

Obtaining usage information

Type a Spatial Analyst function, select it, then right-click and click Usage to see Map Algebra syntax.

See Also

Click About Building Expressions to access help for the Raster Calculator, including the Spatial Analyst functional reference.

Using the Raster Calculator to perform Spatial Analyst functions

1. Click the Spatial Analyst dropdown arrow and click Raster Calculator.
2. Type in the Map Algebra function—for example, slice.
3. Type an open bracket.
4. Double-click the layer you want to use as an input dataset.
5. Type a close bracket or type a comma, add other parameters, then close the brackets.
6. Click Evaluate.

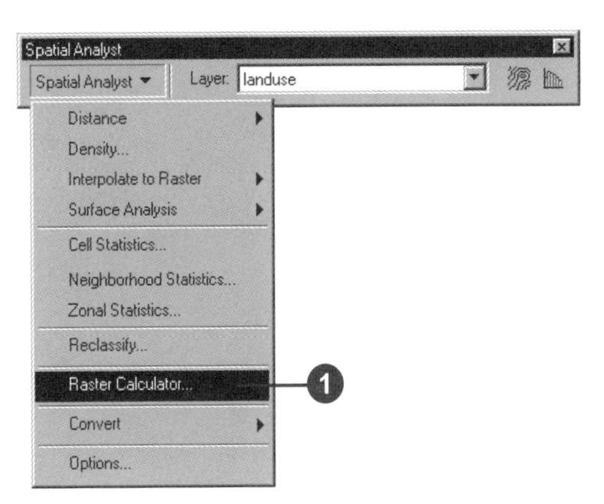

Converting from features to raster

Polygons, polylines, and points from any type of source file can be converted to a raster. It doesn't matter if the source data for the features comes from a CAD drawing, coverage, or shapefile; it can be converted to a raster.

You can convert features using both string and numeric fields. If you use a string field, then each unique string in the field is assigned a unique value in the output raster. A field will be added to the table of the output raster to hold the original string value from the features.

Polygon features to raster

When you convert polygons, cells are given the value of the polygon found at the center of each cell.

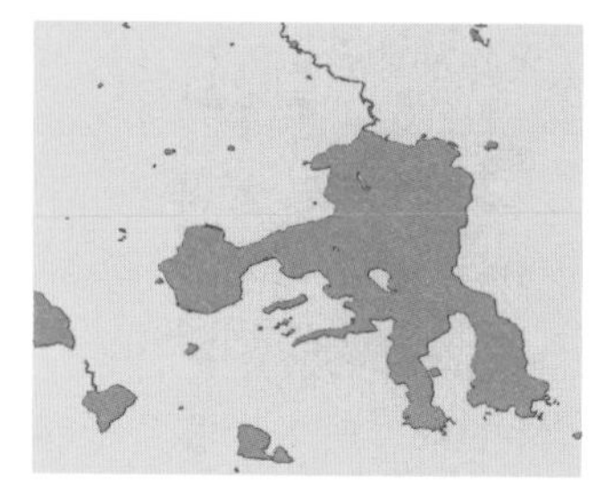

Input polygons

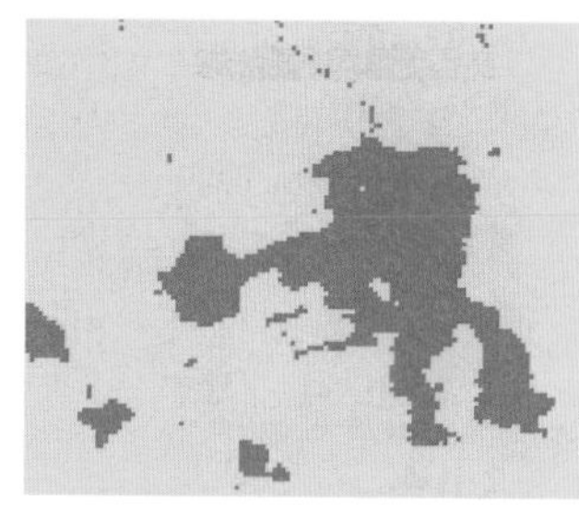

Output raster

Polyline features to raster

When you convert polylines, cells are given the value of the line that intersects each cell. Cells that are not intersected by a line are given the value of NoData. If more than one line is found in a cell, the cell is given the value of the first line it encounters when processing. Using a smaller cell size during conversion will alleviate this.

Input polylines

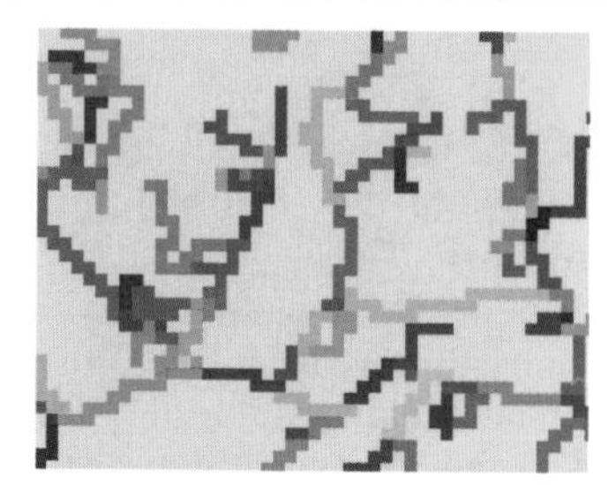

Output raster

Point features to raster

When you convert points, cells are given the value of the points found within each cell. Cells that do not contain a point are given the value of NoData.

If more than one point is found in a cell, the cell is given the value of the first point it encounters when processing. Using a smaller cell size during conversion will alleviate this.

Input points

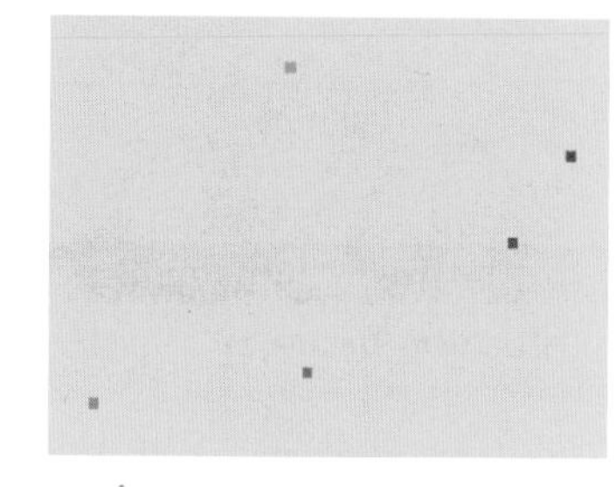

Output raster

Converting from raster to features

Raster to polygons

When you convert a raster representing polygonal features to polygon features, the polygons are built from groups of contiguous cells having the same cell values.

Arcs are created from cell borders in the raster. Continuous cells with the same value are grouped together to form polygons. Cells that are NoData in the input raster will not become features in the output polygon feature.

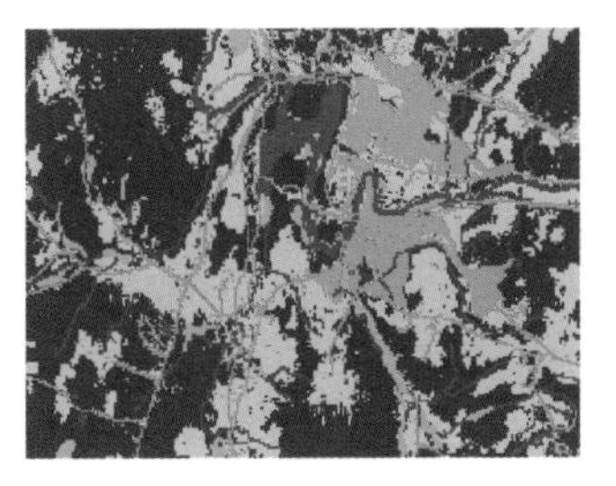

Input raster

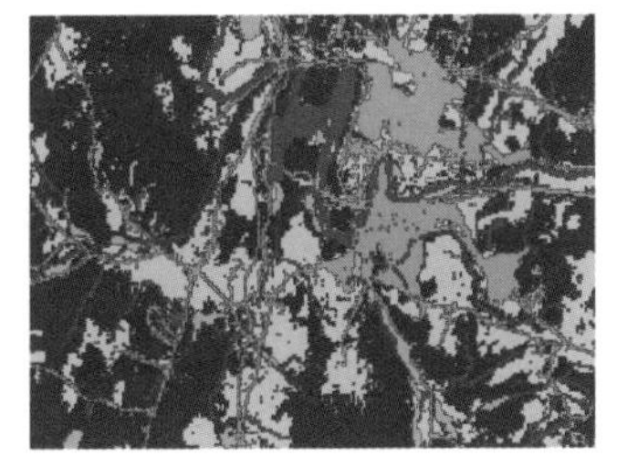

Output polygons

Raster to polylines

When you convert a raster representing linear features to polyline features, a polyline is created from each cell in the input raster, passing through the center of each cell. Cells that are NoData in the input raster will not become features in the output polyline feature.

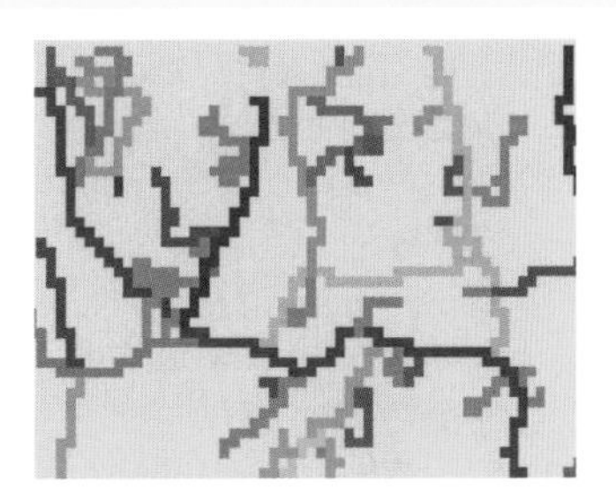

Input raster

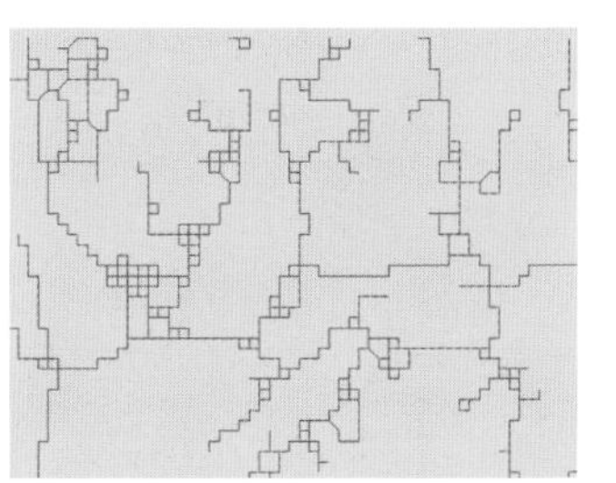

Output polylines

Raster to points

When you convert a raster representing point features to point features, for each cell of the input raster, a point will be created in the output. Each point will be positioned at the center of cell that it represents. NoData cells will not be transformed into points.

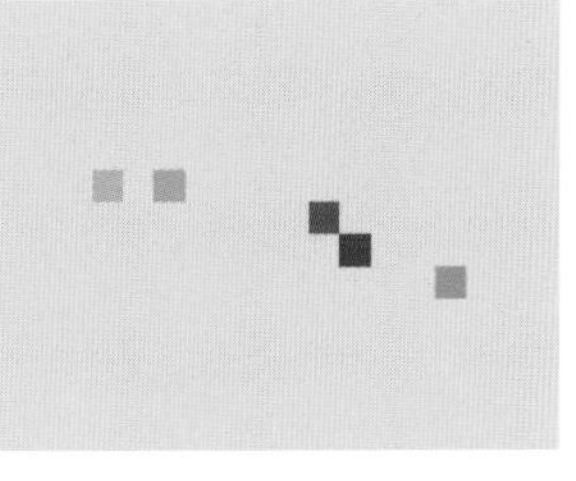

Input raster

Output points

Converting your data

Data can be converted from raster to features or from features to raster.

When converting feature data to a raster, you have the option to specify a cell size for your output raster. The cell size you choose should be based on several factors: the resolution of the input data, the output resolution needed to perform your analysis, and the need to maintain a rapid processing speed. Larger rasters will require longer processing times. A fine resolution, and thus slower processing speeds, may sometimes be essential to your analysis. The default cell size is the smaller of the width or height of the input features extent divided by 250, unless you specified a cell size in the Options dialog box.

Tip

Setting analysis options

Click Options on the Spatial Analyst toolbar to set up your working directory, extent, and cell size for your analysis results.

Converting feature data to raster

1. Click the Spatial Analyst dropdown arrow, point to Convert, and click Features to Raster.
2. Click the Input features dropdown arrow and click the feature you want to convert to a raster.
3. Click the Field dropdown arrow and click the field you want to copy to the Output raster.
4. Optionally, type an Output cell size.
5. Specify a name for the Output raster or leave the default to create a temporary dataset in your working directory.
6. Click OK.

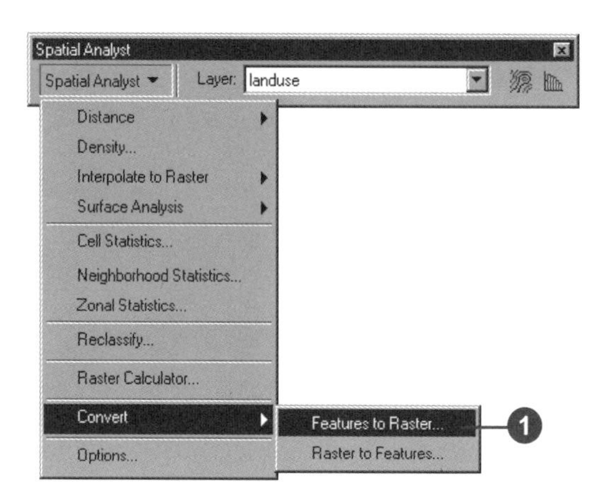

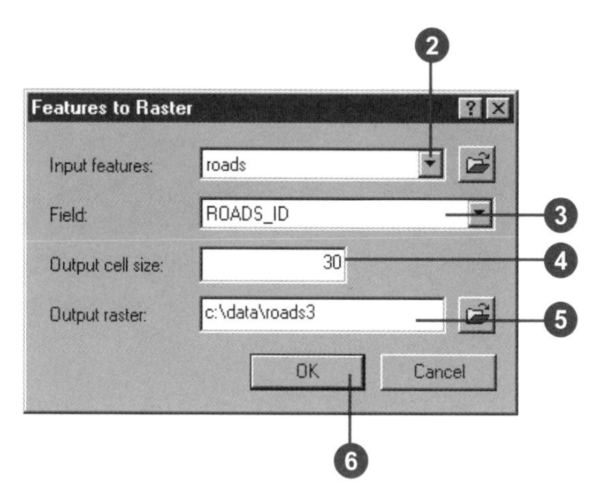

Tip

Browsing for files or directories

If the file you need is not in your table of contents, or if you need to check the directory to place your results, click the Browse button.

Converting raster data to features

1. Click the Spatial Analyst dropdown arrow, point to Convert, and click Raster to Features.
2. Click the Input raster dropdown arrow and click the raster you want to convert to a feature.
3. Click the Field dropdown arrow and click the field with which you want to define features in the Input raster.
4. Click the Output geometry type dropdown arrow and click the type of feature you want to create from your raster data.
5. Specify a name for Output features or leave the default to create a temporary dataset in your working directory.
6. Click OK.

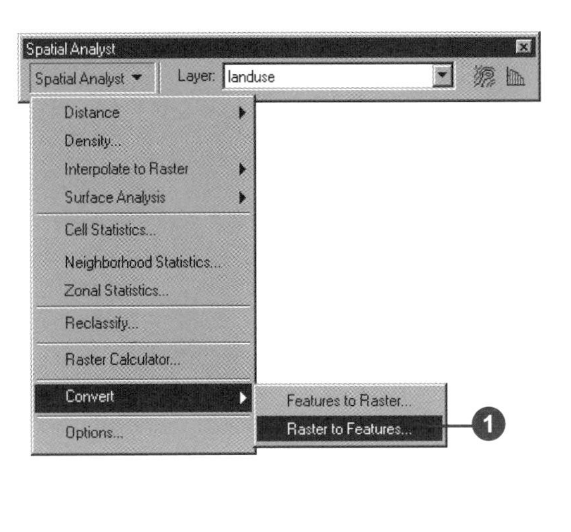

Appendix A

IN THIS APPENDIX

- **Map Algebra language components**
- **Map Algebra rules**

In addition to the many functions that are available through the Spatial Analyst user interface, a wide variety of additional functions are available through Map Algebra, including Spatial Analyst functions. You can access Map Algebra through the Raster Calculator dialog box. Map Algebra expressions can be constructed using the buttons of the Raster Calculator, or they can be typed into the expression box. The expressions will be processed when you click Evaluate.

Map Algebra is the analysis language for Spatial Analyst. It is a simple syntax that is similar to any algebra. Output data will result from some manipulation of the input. The input can be as simple as a single grid dataset, raster layer, or shapefile, and the manipulation can be calculating the sine of each of the location's values, or there can be a series of input grid datasets or raster layers that the manipulation is applied to, such as when adding three grid datasets or raster layers together. Not only does the algebra allow access to additional functions not available in the user interface, but it also allows you to build more complex expressions and process them as a single command. For instance, you can calculate the sine of an input grid dataset or raster layer and add that to two other input grid datasets or raster layers.

Like all languages, Map Algebra is composed of a series of rules. By nderstanding these basic rules, you will be able to use Spatial Analyst in new ways. This appendix outlines the syntax of Map Algebra.

Map Algebra language components

The major strength of Spatial Analyst is its analytical capabilities. Spatial Analyst, through the Map Algebra language, provides tools to perform operations as well as local, focal, zonal, global, and application functions.

The language components

The Map Algebra language provides building blocks that can be used singularly or in conjunction with one another to solve problems. When combining the blocks, a syntax or set of rules must be followed for Spatial Analyst to perform the requested task. The grammar of the language establishes the meaning of the building blocks according to the position of a block in an expression. If type constraints or syntax rules are violated, an error message will be returned by Spatial Analyst and no result will be created.

The building blocks for the Map Algebra language are objects, actions, and qualifiers on the actions. These delineations are similar to nouns, verbs, and adverbs.

Objects

Objects either store information or are values. They are inputs for computation or can be storage locations for output. Grid datasets, raster layers, tables, constants, and numbers are all objects in the Map Algebra language. Any word used in an expression that is not an operator, function, or constant is considered the proper name of an existing or new grid dataset or existing raster layer or table. The function or operator being used provides the context in which to determine the object types.

Actions

Actions that can be performed on input objects are operators and functions. Spatial Analyst operators perform mathematical computations within and between grid datasets, raster layers, tables, and numbers and between valid combinations of them all.

The set of operators is composed of arithmetical, relational, Boolean, bitwise, and logical operators that support both integer and floating-point values and combinatorial operators, which simultaneously overlay grid datasets or raster layers and maintain the input attributes.

Spatial Analyst functions are spatial cartographic modeling tools that analyze cell-based data. These functions are divided into five main categories: local, focal, zonal, global, and application specific. Local functions include trigonometric, exponential, reclassification, selection, and statistical functions. The focal functions provide a set of tools for neighborhood analysis. The zonal functions allow for zonal analysis and computing zonal statistics. The global functions provide tools for full raster layer or grid dataset analysis, such as the generation of Straight Line—Euclidean—and Cost Weighted Distance rasters. The application functions provide tools that are applicable to specific tasks, such as hydrology, data cleanup, and geometric transformation.

Qualifiers

Qualifiers are parameters that control how and where an action is to take place. Even though operators and functions perform actions, the type and manner of the actions vary.

Actions either allow or require qualifying parameters to identify how, to what extent, and with which values the actions are to take place. Which grid dataset or raster layer values should be used in a zonal function, which cells should be included in a focal neighborhood, and what power to raise the input values in a power function are examples of parameters necessary for the completion of a Spatial Analyst action.

Constants and numbers are single-value objects, usually numerical, that can be used in conjunction with an operator or function to achieve a desired result. Some of the built-in constants available in the Map Algebra syntax and language are

PI (3.14), E (2.718), and DEG (57.296)—degree/radian. All of the values of a grid dataset or raster layer can be multiplied or divided by any number, or a number can be added to or subtracted from each value in a grid dataset or raster layer. Numbers can be used in most operations on a grid dataset, raster layer, or constant. When used in a function, a number can also set a parameter, such as a neighborhood width, the maximum distance to which to calculate the Straight Line—Euclidean—distance, or the test for a conditional statement.

Map Algebra syntax

Operators can be placed between one or two input grid datasets, raster layers, numbers, or constants.

```
[inlayer1] + [inlayer2]
```

In the above equation, an output raster dataset is created storing the results from an expression, adding the values for the raster layers inlayer1 and inlayer2 on a cell-by-cell basis.

Or the operator can be placed in front of the input grid datasets, raster layers, numbers, or constants.

```
- [inlayer1]
```

In the above equation, an output raster dataset is created where each cell location will contain the negative of the value of the corresponding cell location in the raster layer inlayer1.

An expression performing a function is dependent on the syntax and parameters associated with each function.

```
sin(c:\data\ingrid1)
mean([inlayer1], [inlayer2], [inlayer3])
focalsum([inlayer1], rectangle, 3, 3)
zonalmean([inlayer1], c:\spatial\ingrid2)
eucdistance(e:\data2\ingridsource)
```

All of the previous expressions are valid. In the first expression, an output raster dataset receives the sine of the values for the grid dataset ingrid1 on a cell-by-cell basis. In the second expression, an output raster dataset receives the mean of the values of the raster layers inlayer1, inlayer2, and inlayer3 on a cell-by-cell basis. In the third expression, each cell in the output raster dataset receives the result of the sum of the values of the eight immediate neighbors and the processing cell itself. The fourth expression results are the mean of the values in the grid dataset ingrid2 delineated by the zones of the raster layer inlayer1. The final expression assigns to each cell in the output raster dataset its Straight Line, or Euclidean, distance from the set of source cells in the grid dataset ingridsource.

Compound or nested expressions that perform multiple tasks can also be created. They are formed by combining constants, numbers, grid datasets, raster layers, item names, and tables with operators and functions. Each compound expression can consist of multiple actions, as in:

```
sin([inlayer1]) + pow([inlayer2], 2)
```

Each operator has an assigned precedence value (see 'Table of supported operators and precedence values' in Appendix B). Spatial Analyst processes the operator with the highest precedence first, the next-highest precedence second, and so on. If two operators in an expression have the same precedence value, Spatial Analyst, which reads left to right, will process the operator that is located farthest to the left first. All functions have equal precedence and therefore are processed simply from left to right.

Parentheses overrule precedences, so Spatial Analyst completes all operations within parentheses first. Parentheses can be nested, with the deepest-nested operator or function being processed first.

Spatial Analyst Map Algebra is similar to, and follows many of the conventions of, the standard algebraic order of operations. The main difference between the two is that Spatial Analyst Map Algebra has been designed to work on grid datasets and raster layers, while standard algebra works on numbers.

Map Algebra input types

Valid input types in the Map Algebra language are grid datasets, raster layers, shapefiles, coverages, tables, constants, and numbers.

All input grid datasets, raster layers, shapefiles, and tables must exist prior to processing the expression.

When entering a grid dataset, shapefile, coverage, or table into an expression, the name can be used directly if it resides in the current working directory—set in the Options dialog box—as in the following example:

```
cos(inlayer1)
```

The full pathname to the grid dataset, shapefile, coverage, or table must be identified if it is not in the current workspace and is not a raster layer added to your ArcMap session:

```
cos(E:\mydirectory\ingrid1)
```

If the input is a raster layer in the Layers list in the Raster Calculator, the name of the layer needs to be in brackets.

```
cos([inlayer1])
```

Map Algebra results

The result from a Map Algebra expression can be a raster, a shapefile, a table, or a file stored on disk, such as an ASCII file. The output dataset name does not need to be specified, as a temporary dataset can be created; Spatial Analyst will name the temporary output dataset calc followed by a number—that is, calc1. The number following calc will increase incrementally to the next unique value for the name of each new output dataset.

Any integer raster dataset returned as the result of a function or operator will usually have an associated table—a Value Attribute Table, or VAT—with two default items: Value and Count. Some operators return raster datasets in which the associated table has additional items. For combinatorial operators and functions, the output not only contains the Value and Count of a new raster dataset, but also identifies the combination of values in the input that resulted in each output value.

Multiple outputs

While most functions generate a single new raster dataset as output, a few functions, such as the Euclidean Allocation and Cost Allocation, create multiple output raster datasets. A function that generates multiple output raster datasets returns one of the raster datasets as its primary result. The remaining raster datasets appear as optional output arguments to the function. They will be permanently saved to the current working directory if output raster names are supplied for the arguments or to the directory you specify if a full pathname is supplied.

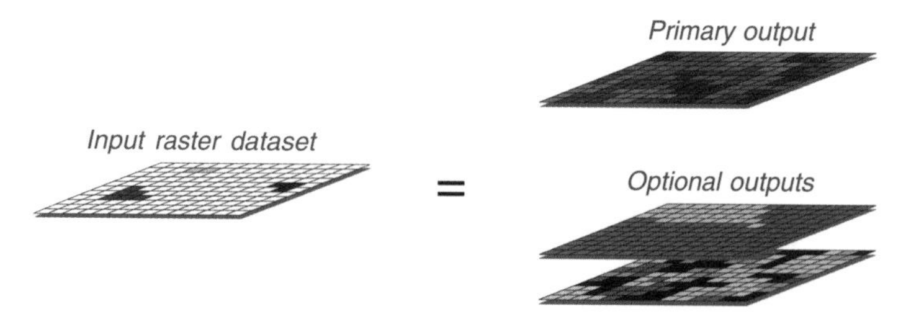

Output value types

The input value type can determine the value type of the resultant raster dataset. Generally, when an operator—not a function—is applied to one or more input integer grid datasets or raster layers, the result will be an integer raster dataset; when an operator is applied to one or more floating-point grid datasets or raster layers, the result will be a floating-point raster dataset. When an operation is performed on two or more grid datasets or raster layers, and at least one of the input grid datasets or raster layers contains floating-point values, a floating-point raster dataset will result. There are exceptions to these rules. The Boolean and combinatorial operators, for example, always output integer values, no matter what the input types are.

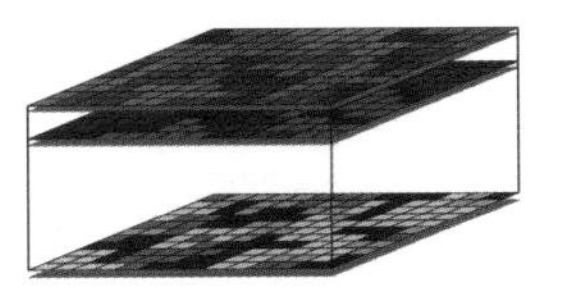

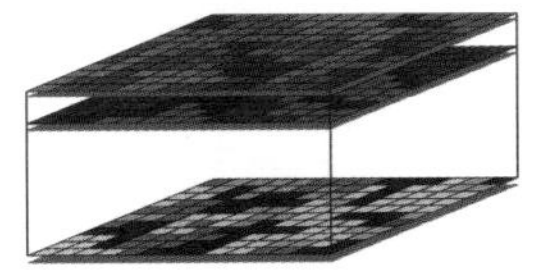

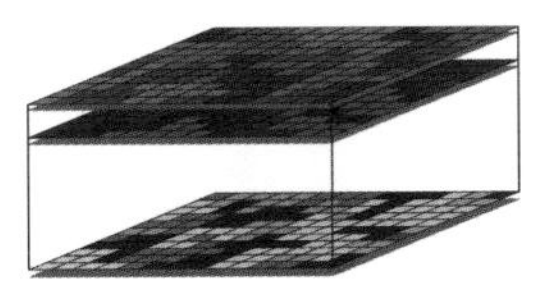

Floating-point values are returned by all functions that perform statistical calculations, such as the mean and standard deviation for local, focal, and zonal functions. Some global functions, such as the distance and interpolation functions, return floating-point results. With other functions, such as Select, FocalSum, and ZonalMin, the input value type dictates the output value type. The output type is specifically listed with each command in the ArcGIS Desktop Help system.

Map Algebra provides several functions for converting between floating-point and integer raster types. For additional information, refer to the ArcGIS Desktop Help system discussions of Int, Float, Floor, and Ceil.

NoData in operators and functions

The general behavior of NoData in Spatial Analyst is as follows:

- For any operator or local function, if any cell location of any of the input grid datasets or raster layers is assigned NoData, the output for the cell location will be NoData.
- For focal functions, if any cell location in a neighborhood of a processing cell is assigned NoData, the function will ignore the NoData value and compute with the remaining values. The keyword NoData can be used to override this default behavior to return NoData as the output for the processing cell location when any cell within the neighborhood contains NoData.
- For zonal functions, if any cell location on the input value grid dataset or raster layer in a zone defined by the input zone grid dataset or raster layer is assigned NoData, by default the function will ignore the NoData value and compute with the remaining values. The keyword NoData can be used to override this behavior and return NoData to the cell location when any cell within the zone contains NoData. If NoData exists for any cell location on the input zone grid dataset or raster layer, the output value for the location is NoData.
- For straight line—Euclidean—distance functions, the NoData value is ignored for computations since the distance and

direction are true Straight Line distance and direction. The input source grid dataset or raster layer must contain valid values for source cells and NoData for nonsource cells.

- For cost-distance functions, any cell assigned NoData in the cost grid dataset or raster layer will be considered a barrier in computations, and the cell locations containing NoData values on the input cost grid dataset or raster layer will contain NoData on the output. The input source grid dataset or raster layer must have valid values for source cells and NoData for nonsource cells.
- For the remaining global functions, when NoData exists at any location on the input grid dataset or raster layer, the output value for that location will receive NoData.
- With the Select function, when the results of the evaluation of the conditions are not True in the expression, NoData is returned, rather than 0. This is in contrast to the Relational operators and the Test function, which will return 0 for those cells for which the results of the evaluation are False.
- For the conditional function, Con, if no value is assigned to the false-expression output argument, which is used when the result of the evaluation of the condition is False, cells that evaluate to False will receive NoData on the output.
- Some local functions, such as Popularity, Majority, and Minority, evaluate the number of occurrences of a value, not the value itself. If there is no single value found to be the *nth* most popular, minority, or majority, NoData will be assigned to the output for the location. This situation occurs when all input values at a location are different. No value ever occurs the majority or the minority of times. Returning one of the input values, such as the first one encountered, would be incorrect. You would not know whether the value is truly a majority, minority, or the *nth* most popular value.

Each command reference in the Spatial Analyst ArcGIS Desktop Help system contains the specific handling of NoData in the computations.

Assigning values to NoData and NoData to values

At times, it may be desirable to replace cells with NoData in a grid dataset or raster layer with some other valid value. The desired result for the output of the expression is not to treat the NoData values as NoData, but to treat them as zeros, or some other value. There are many ways to change the NoData assignments to valid values in Spatial Analyst. From the user interface, use the Reclassify dialog box, or with the Map Algebra, use the IsNull and Con functions.

```
con(isnull ([inlayer1]), 0, [inlayer1])
```

The above expression says if—Con—the cell value on inlayer1 equals NoData—IsNull—then assign 0 to it (true_expression = 0); if it does not equal NoData—it is a valid value—assign the value of inlayer1 (false_expression = inlayer1).

To perform the reverse and set cells with specific values to NoData—to mask out cells—use the SetNull function.

```
setnull([inlayer1] > 100, [inlayer1])
```

The above expression will assign all values greater than 100 to NoData. The cell locations currently with NoData will remain NoData, and the remaining cell locations will retain their input values.

Locations outside the grid dataset or raster layer limits or beyond the analysis extent are considered to be NoData.

Conditional statements

The Con function

The Con function's name is short for conditional statement. The Con function is a local function that is evaluated on a cell-by-cell basis. The usage of the Con function is:

```
Con (condition, true_expression, condition,
true_expression, condition, true_expression,
false_expression)
```

where condition is a conditional expression that is evaluated for each cell in the analysis extent. If the condition is True, true_expression identifies the value used to compute the output cell value. Additional condition statements can be tested on the values of the input grid dataset or raster layer with a mandatory true_expression identifying the value to be applied to those cells where the additional statement is True. If none of the results of the evaluations of the conditional statements are True, a value or expression can be applied to the cells through the false_expression optional argument. If no value is specified at the false_expression argument, any cell that does not meet any of the conditions within the expression will be set to NoData. An example of a simple Con function is:

```
con([inlayer1] > 5, 10, 100)
```

In the above expression, if the value of a cell location in the raster layer inlayer1 is greater than 5, 10 will be assigned to that cell location on the output raster dataset, while locations on inlayer1 with values 5 or lower will be assigned 100 on the output raster dataset.

If no value or expression is specified for the False expressions,

```
con([inlayer1] > 5, 10)
```

the results will be the same as the above output except that the cells that have a value of 5 or less in the raster layer inlayer1 will be assigned NoData on the output raster dataset.

Any valid expression (see 'Map Algebra rules' later in this appendix) can be used in place of a value for the true_expression and false_expression arguments.

```
con([inlayer1] > 5, sin([inlayer1]),
cos([inlayer1]))
```

In the above expression, the sine of all values greater than 5 and the cosine for all values of 5 or less are calculated, and the results are sent to an output raster dataset.

Multiple conditional statements can be used within the Con function, but each must have a value or expression for true_expression that can be used to assign values to the output cells if the result of the evaluation of the condition is True. The optional value or expression false_expression can be applied if none of the results of the evaluations of the conditions are True.

```
con([inlayer1] < 5, sin([inlayer1]), [inlayer1] <
20, cos([inlayer1]), [inlayer1] > 50, 100, 0)
```

In the above expression, the sine is calculated for those values that are less than 5; the cosine is calculated for the values that are 5 or greater but less than 20; the values greater than 50 are assigned 100; and the remaining values, which are 20 or greater but less than 50, are assigned 0.

Multiple parameters can be used in a conditional expression of the Con function.

```
con(([inlayer1] > 5 & [inlayer1] < 10), 5, 100)
```

Operators and functions can be applied to the input grid datasets and raster layers in the conditional expression and the results evaluated.

```
con(sin([inlayer1]) > .5, 10, 100)
con(([inlayer1] + [inlayer2]) > 10, 100, 5)
```

```
con([inlayer1] > 5, cos([inlayer1]),
sin([inlayer1]))
```

A Con function can be nested within another Con function.

```
con([inlayer1] > 23, 5, con([inlayer1] > 20, 12,
con((([inlayer1] > 2) & (ingrid1 < 17)),
sin([inlayer1]), 100)))
```

Multiple grid datasets and raster layers can be used in the conditional statement or in the expression to be performed on the cells.

```
con([inlayer1] + c:\data\ingrid2 > 7,
sin([inlayer1]), cos(c:\data\ingrid2))
```

```
con([inlayer1] < 9, [inlayer1] * c:\data\ingrid2
+ tan([inlayer3]), cos([inlayer1]))
```

Map Algebra rules

The following notes are a quick reference to using Spatial Analyst Map Algebra. By stating rules and providing examples, an overview of Map Algebra is presented. This section only presents the grammar of the language. Examples may not replicate your exact expression, but by breaking down the components of your expression, you should be able to find the grammatical rules that apply to its pieces.

General Map Algebra rules

The result of a Map Algebra expression in the Raster Calculator can be a raster dataset, a shapefile, a table or a file stored on disk such as an ASCII file.

All operators must be separated from their operands by blank spaces on both sides:

```
[inlayer1] * [inlayer2] div c:\data\ingrid3
[inlayer1] & [inlayer2]
[inlayer1] + c:\results\ingrid2 - [inlayer3]
```

Parentheses are not operators and do not need the blank spaces on both sides:

```
([inlayer1] div [inlayer2]) * [inlayer3]
[inlayer1] + ([inlayer2] + 8)
(([inlayer1] * 6) + [inlayer2]) & d:\data\ingrid3
```

Grid datasets, raster layers, shapefiles, coverages, tables, and item names can consist of most combinations of characters and numbers:

```
[inlayer_1] + [inlayer2]
[inlayer12345] + [inlayer2]
```

Symbols such as (, {, and \ cannot be used in a name.

Map Algebra rules for operators

Most operators can be utilized on multiple integer or floating-point grid datasets or raster layers:

```
[inlayer1] * [inlayer2]
[inlayer1] && [inlayer2]
[inlayer1] diff [inlayer2]
```

Generally, the operator is placed between two input grid datasets or raster layers; however, due to the nature of the operator, unary operators are placed before a single input grid dataset or raster layer:

```
- [inlayer1]
^^ c:\mydirectory\ingrid1
^ [inlayer1]
```

Whenever the NoData value is encountered at a cell location on any of the input grid datasets or raster layers, no matter what the operator, the output for that location will receive NoData as a result of the operation.

Multiple operators can be used in building an expression:

```
[inlayer1] + [inlayer2] - [inlayer3]
e:\sp\ingrid1 mod [inlayer2] div e:\sp\ingrid3
^ [inlayer1] & [inlayer2]
```

Parentheses and expressions with multiple operators, grid datasets, and raster layers

When multiple operators are used in an expression, the order of processing is dependent on the precedence value assigned to the operator (see Appendix B). The higher the precedence value, the sooner the operation will be processed.

When two operators in the same expression have the same precedence value, the one that is farthest to the left is processed first:

```
[inlayer1] + c:\input_data\ingrid2 - [inlayer3]
```

Precedence values can be overridden using parentheses. The expression within the innermost parentheses is processed first, no matter what the precedence value of the operator:

```
([inlayer1] diff [inlayer2]) * [inlayer3]
```

```
[inlayer1] + (c:\data\ingrid2 & [inlayer3])
```

```
[inlayer1] / ([inlayer2] - [inlayer3])
```

When two or more sets of parentheses at the same level are used, each set has the same precedence value; therefore, the expression in the set of parentheses farthest to the left is processed first:

```
([inlayer1] + [inlayer2]) / ([inlayer3] !!
[inlayer4])
```

```
([inlayer1] - [inlayer2]) >> ([inlayer3] mod
c:\spatial\ingrid4)
```

Parentheses can be nested. The expression within the innermost parentheses will be processed first:

```
([inlayer1] + ([inlayer2] - ([inlayer3] >>
[inlayer4]))) / [inlayer5]
```

```
[inlayer1] >> (( ^ [inlayer2]) ! [inlayer3])
```

```
([inlayer1] diff ([inlayer2] - ([inlayer3] &&
([inlayer4] - [inlayer5])))) div [inlayer6]
```

Operators with numbers

Numbers can also be used in the expressions:

```
[inlayer1] + 5
```

```
c:\data\ingrid1 > 8
```

```
[inlayer1] diff 3
```

Sometimes numbers are used as parameters within an expression involving an operator:

```
[inlayer] in {0, 3, 5, 8}
```

Expressions need not contain any data and can be constructed using only numbers and operators. The output dataset will default to the existing dataset size and cell resolution that have been set in the analysis environment:

```
5
```

The output will be a raster where each cell value contains the value 5.

```
9 + 20
```

The output will be a raster where each cell value contains the value 29.

Operations with numbers and rasters

Numbers can be used in the creation of compound expressions:

```
[inlayer1] / [inlayer2] + 5
```

```
[inlayer1] < 2 * 35
```

```
[inlayer1] <= 40 - [inlayer2] + 7
```

The order of processing is still dependent on the precedence value assigned to the operator. The order of processing can be overridden with parentheses. All of the rules for the precedence value and parentheses apply when expressions mix grid datasets, raster layers, numbers, and operators:

```
([inlayer1] + 5) * 20
```

```
[inlayer1] / [inlayer2] - (5 - 2)
```

```
10 * ([inlayer1] + (6 / ([inlayer2] diff
[inlayer3])))
```

Map Algebra rules for functions

All functions begin with the function name, followed by the grid dataset, grid datasets, raster layer, or raster layers to which the function is to be applied and the necessary parameters, all in parentheses:

```
tan([inlayer])
focalmax([inlayer1], rectangle, 4, 4)
zonalmin([zonelayer], [valuelayer])
```

The arguments or parameters within functions are separated by commas:

```
focalmin([inlayer1], circle, 6)
zonalmax([zonelayer], c:\data\valuegrid)
```

Many functions also have additional parameters. These parameters may be keywords, numbers, names of tables, and even other rasters. The parameters are function dependent:

```
selectbox([inlayer1], 45, 67, 200, 360)
focalrange([inlayer1], annulus, 2, 4)
zonalmean([zonelayer], c:\data\valgrid, NoData)
```

Compound expressions

Functions can be used in compound expressions together with operators, grid datasets, raster layers, shapefiles, coverages, and numbers:

```
sin([inlayer1]) + [inlayer2]
focalsum([inlayer1], rectangle, 3, 3) *
tan([inlayer2])
zonalmin([zonelayer], [valuelayer]) - 3
```

All functions have the same precedence value; therefore, when multiple functions are used in an expression, they are evaluated from left to right:

```
min([inlayer1], [inlayer2], [inlayer3]) +
abs([inlayer4])
ceil([inlayer1]) * slice([inlayer2], eqarea, 10)
popularity(2, [inlayer1], [inlayer2], [inlayer3])
* tan([inlayer4])
```

All rules that apply to parentheses for expressions built with operators, grid datasets, and raster layers apply to functions within expressions. The function or operator that is within the most deeply nested parentheses will be processed first. As for an operator, the output of a function is a raster dataset, and that raster dataset can be used as further input in the expression:

```
(sin([inlayer1]) +
focalrange(c:\data\ingrid2,circle, 7)) - 6
[inlayer1] * (zonalmax([inlayer2] +
e:\algebra\ingrid3) + [valuelayer] / 8)
(majority([inlayer1], ([inlayer2] - [inlayer3]),
[inlayer4]) && (^ ([inlayer5] - 10)) /
[inlayer6]) > 8
```

Functions can be composed as long as the output of the inner function is of the same type as the corresponding argument to the outer function:

```
sin(focalmean ([inlayer1])) < 2 * 3
regiongroup(reclass([inlayer1],
c:\data\reclass_table.txt))
majority(([inlayer1] + [inlayer2]),
cos([inlayer3]), zonalmin([inlayer4],
[inlayer5]))
```

Appendix

IN THIS APPENDIX

- **Table of supported operators and precedence values**
- **About precedence values**

The Raster Calculator provides the ability to use a full suite of operators to perform analysis within and between multiple rasters. This section provides a table, which lists all supported operators, followed by a discussion on precedence values. A brief description accompanies each operator in the table, followed by its precedence value. For more information on the operators, refer to 'The Raster Calculator' in Chapter 7.

Table of supported operators and precedence values

The table below lists all supported operators. A brief description accompanies each operator, followed by a precedence value.

	OPERATORS	
Operator	**Description of operator**	**Precedence**
Arithmetic:		
-	unary minus	12
mod	modulus	11
*	multiplication	11
/	division	11
div	Floating-point division	11
+	addition	10
-	subtracts	10
Boolean:		
^, not	complement of expression	12
&, and	and	3
!, or	exclusive or	2
l, xor	or	2
Relational:		
<, lt	less than	6
<=, le	less than and equal to	6
>, gt	greater than	6
>=, ge	greater than and equal to	6
==, eq	equal to	6
^=, <>, ne	not equal to	6
Bitwise:		
^^	bitwise complement of expression	12
>>	right shift	7
<<	left shift	7
&&	bitwise and	5
!!	bitwise exclusive or	4
ll	bitwise or	4
Combinatorial:		
cand	and	9
cor	or	8
cxor	exclusive or	8
Logical:		
diff	logical difference	8
in	contained in	8
over	over	8

About precedence values

The precedence value determines the priority of processing for each operator. The larger the precedence number assigned in an operator, the higher its priority, and thus the sooner the Spatial Analyst interpreter will process it. The Spatial Analyst interpreter processes the operator with the highest priority first, the second-highest priority second, and so on.

```
- [inlayer1] + [inlayer2] div [inlayer3]
```

In the above expression, the negative of inlayer1 is first calculated, then inlayer2 is divided by inlayer3 and, finally, the two temporary resultant rasters are added together (the result of - inlayer1 + the result of inlayer2 div inlayer3).

Operators with the same precedence level are processed from left to right.

```
[inlayer1] * [inlayer2] div [inlayer3]
```

To process the above expression, the Spatial Analyst interpreter will first multiply inlayer1 with inlayer2, then divide the result by inlayer3.

The left-to-right rule applies to all operators except for shift operators, << and >>, with precedence level 7 and unary operators, unary -, ^, and ^^, with precedence level 12. These two operators have a right-to-left association.

```
- ^ ^^ [inlayer1]
```

First the bitwise complement, ^^, of inlayer1 is calculated, then the Boolean complement, ^, is taken from the results of the bitwise complement and, finally, the unary minus is determined from the previous resultant raster.

Appendix C

IN THIS APPENDIX

- **About remap tables**
- **Slice and remap tables**
- **Reclass and remap tables**
- **Slice versus Reclass relative to remap tables**

The Reclassify function in the Spatial Analyst user interface enables you to quickly and easily reclassify your data and save the reclassification table if you wish for later use. The format of this table is such that it allows the mapping of NoData to a value; mapping a value, range, or string to NoData; or mapping strings to new values.

As an alternative, remap tables (INFO™ and ASCII) can be used in the Raster Calculator through the Reclass and Slice functions. This chapter explains the rules for creating these INFO and ASCII remap tables, giving examples of their use in the Reclass and Slice functions.

About remap tables

Overview of reclassification

Remap tables can be applied to rasters via the Raster Calculator using either the Reclass or the Slice function. However, you do not need to use remap tables to reclassify your data using Spatial Analyst. You can simply use the Reclassify function on the Spatial Analyst toolbar to reclassify your data and save the table you create if you wish. Doing so will enable you to load it again at a later date.

Remap tables described here cannot be used in the Reclassify dialog box.

Remap tables

Remap tables can be either ASCII files or INFO tables. They consist of two parts. The first part identifies the particular cell value to be reclassified, and the second part is the cell's reclassified output value.

INFO remap tables

Value	Symbol
3	1
5	2
10	3
15	4

Cells with a value less than or equal to 3 are assigned symbol 1.

Cells with a value greater than 3 and less than or equal to 5 are assigned symbol 2.

Cells with a value greater than 5 and less than or equal to 10 are assigned symbol 3.

Cells with a value greater than 10 and less than or equal to 15 are assigned symbol 4.

Cells with a value greater than 15 are assigned NoData.

ASCII remap tables

The ASCII table functions in the same manner as the INFO counterpart but allows for much greater flexibility in determining the reclassified values. The remap table can be created with any text editor using the formatting rules discussed in the following paragraphs to define the parameters for reclassification.

The ASCII remap table is made up of optional comments, optional keywords, and required assignment statements. Each statement must be on a separate line. Comments are descriptive text that can be used to provide any additional information that needs to be included. Comments can appear anywhere in the remap table but must be preceded by a pound sign (#). The keywords establish the parameters in which the reclassification operates. The assignment statements assign an output value to a specified input cell value or range of values.

The keywords are positioned at the beginning of the file before any assignment statements are entered. Comments, however, can be anywhere and may precede the keywords. There are two optional keywords that can be included in the lookup table. The first is LOWEST-INPUT, which identifies the lowest cell value in a raster to consider for reclassification. LOWEST-INPUT is formatted as follows: lowest-input <value>, where <value> is the minimum cell value to consider for reclassification. LOWEST-INPUT is used when you want to exclude cells with values below the specified value. For example, in a raster with cell values ranging from one to 20, setting LOWEST-INPUT to 5 would exclude all those cells with a value less than 5. If not specified, LOWEST-INPUT defaults to the minimum value in the input raster.

The second optional keyword, LOWEST-OUTPUT, identifies the lowest output value or starting point for the reclassified values. This keyword is used to set the output reclassified values automatically for cases where the assignment statements, described later in this appendix, specify only an input value. LOWEST-OUTPUT is formatted: lowest-output <value>, where <value> is the lowest output reclassified value. If not specified, LOWEST-OUTPUT defaults to 1.

The assignment statements follow the keywords. They can be formatted using several different methods. The general form of an assignment statement establishes the relationship between an input cell value and its reclassified value:

input cell value ⟶ output reclassified value

The input cell value can be either an integer or a real number. The output reclassified value, however, can only be an integer.

Several methods can be used to specify an input value and its associated reclassified value. These methods are best presented with examples. The remaining discussion on ASCII remap tables presents a remap table and describes how the input cell values are reclassified according to the table. All examples use a raster dataset with cell values from one to 20. The first example shows a remap table with assignment statements that contain only an input cell value.

```
# Example 1
# Remap table for cell value reclassification
LOWEST-INPUT 3
LOWEST-OUTPUT 2
5
6
7
15
```

Input cell values must always be sorted in ascending order.

As with an INFO remap table, the successive assignment statements implicitly define ranges of cell values for reclassification. Thus, it is essential that the input cell values be sorted in ascending order. The output reclassified value for each range is automatically calculated from the value specified with LOWEST-OUTPUT. The first range of cell values is reclassified to the value specified with LOWEST-OUTPUT. The next range is reclassified to LOWEST-OUTPUT plus one, and so on, until all assignment statements have a reclassified value. Any cell values that fall outside the specified ranges are reclassified to NoData.

The following table summarizes the reclassification:

Input Cell Value	Output Reclassified Value
Less than 3	NoData
3 to 5	2 (lowest-output)
Greater than 5 to 6	3 (lowest-output + 1)
Greater than 6 to 7	4 (lowest-output + 2)
Greater than 7 to 15	5 (lowest-output + 3)
Greater than 15	NoData

If LOWEST-INPUT has not been specified, all cell values less than or equal to 5 would have been reclassified to 2. The reclassified value would have defaulted to 1 if a LOWEST-OUTPUT of 2 was not specified.

The first method shows how reclassification can be limited to those cell values that fall between a minimum and a maximum value. It does not, however, provide control over the cell values within the minimum and maximum. To obtain this kind of control, explicit ranges of input values can be specified. For example:

Example 2

Remap table for cell value reclassification

LOWEST-OUTPUT 2

3 5

5 9

13 15

With this method, LOWEST-INPUT is ignored. LOWEST-OUTPUT automatically causes the reclassified values to be generated for the input ranges. Remember that the ranges must be sorted in ascending order. They also should not overlap except at the border of two ranges. Thus, an input range from 5 to 9 followed by an input range from 8 to 12 is not valid. For the above remap table, the input cell values are reclassified as shown below.

Input Cell Values	Output Reclassified Value
Less than 3	NoData
3 to 5	2
Greater than 5 to 9	3
Greater than 9 to 13	NoData
Greater than 13 to 15	4
Greater than 15	NoData

By omitting an assignment statement for the range for nine to 13, all cells that fall within this range are reclassified as NoData and are displayed with the symbol for NoData.

User-specified output values for each input value or input range can be specified by adding an additional field to the remap table. The input cell value or value range is followed first by a colon (:) and then by the desired output reclassified value. When an explicit output reclassified value is specified, LOWEST-OUTPUT is ignored.

For example:

Example 3

Remap table for cell value reclassification

LOWEST-INPUT 3

5 : 10

6 : 16

7 : 62

15 : 28

The reclassification is summarized in the following table:

Input Cell Values	Output Reclassified Value
Less than 3	NoData
3 to 5	10
Greater than 5 to 6	16
Greater than 6 to 7	62
Greater than 7 to 15	28
Greater than 15	NoData

Similarly, an output value can be specified for explicit input ranges.

Example 4

Remap table for cell value reclassification

3 5 : 9

5 9 : 8

13 15 : 59

The reclassification is summarized as shown below:

Input Cell Values	Output Reclassified Value
Less than 3	NoData
3 to 5	9
Greater than 5 to 9	8
Greater than 9 to 13	NoData
Greater than 13 to 15	59
Greater than 15	NoData

All of the examples presented above are valid ASCII remap tables that can be used to reclassify cell values. Each of the four methods presented above shows the acceptable syntax for an ASCII remap table. The syntax cannot be mixed among the four types. For example, it is not valid to specify an assignment statement that contains a single input value followed by another assignment statement that contains an input range.

This is an invalid remap table. Single input cell values and input ranges cannot be specified in the same remap table.

Invalid remap table for cell value reclassification

LOWEST-INPUT 3

LOWEST-OUTPUT 2

5

6 9

11

15

Also, it is not valid to specify an output reclassified value only on some of the assignment statements in a remap table. If a user-specified output value is entered, it must be specified in all assignment statements.

This is an invalid remap table. All assignment statements must have a specified output value.

Invalid remap table for cell value reclassification

LOWEST-INPUT 3

5 : 10

6

7 : 62

15

Slice and remap tables

Slice uses remap tables to change ranges of values.

Using the Slice function, you can identify the item names in the INFO remap table for the input and output columns.

slice (<raster>, {TABLE}, <remap_table>, {in_item}, {out_item}, {in_min})

Example:

```
slice([inlayer1], table, remap_table, type, code)
```

In the above example, table is a keyword defining the type of slicing, remap_table is the name of the remap table, and code is the name of the output column.

The item containing the values to be sliced and the item containing the output value do not have to be adjacent in the INFO table. If no names are specified for the input and output fields, the default fields are VALUE and LINK. If the specified input or output fields do not exist, or if no fields are specified and the INFO table does not have VALUE and LINK fields, an error will be returned.

Grid.item syntax can be used in the Raster Calculator. In the input Slice expression, if no .item is identified for the input, the values in the input field in the INFO table will be mapped to the VALUE item in the grid's VAT. The input field name in the INFO remap table does not have to be VALUE for this mapping to occur. If a range exists in the INFO remap table that is beyond the values in the VALUE field of the VAT, the range will be ignored. The values associated with the specified VAT item are mapped to the corresponding cell locations. The INFO remap table item is then used to reclassify the cell values.

The pages that follow give a graphical representation of the use of remap tables in the Slice function.

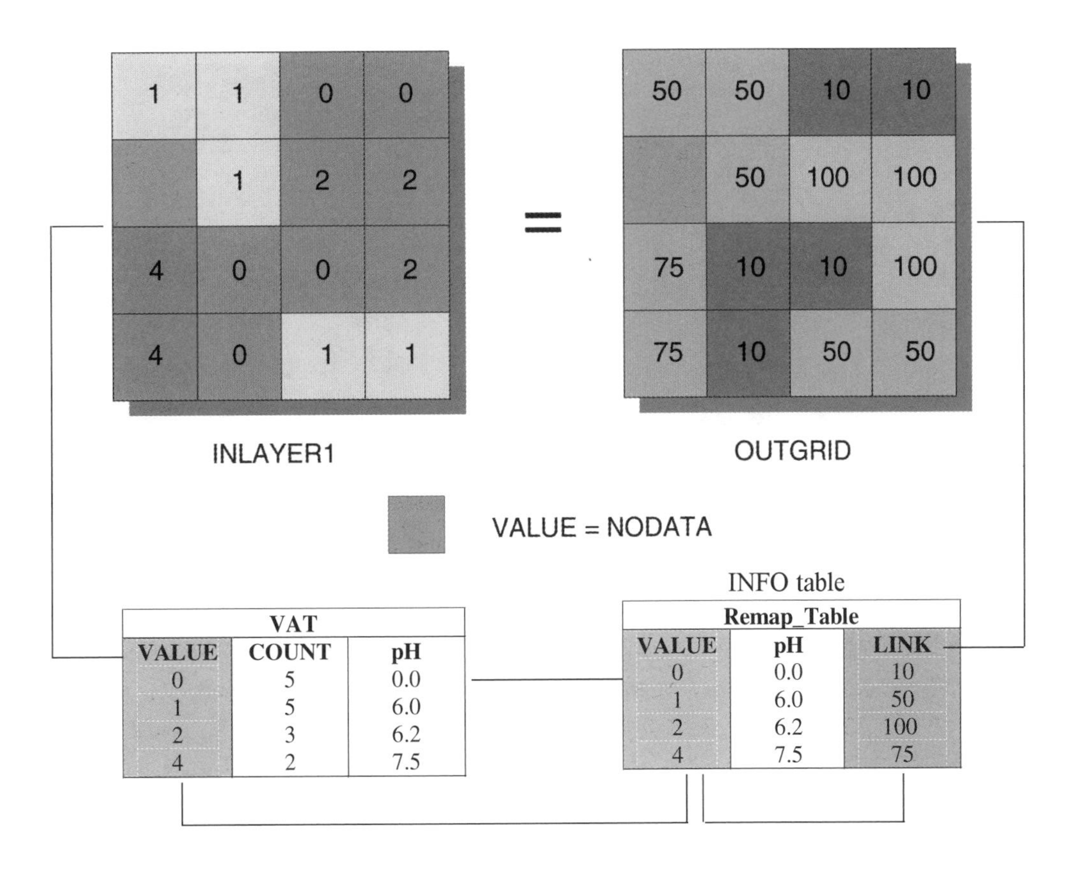

VAT		
VALUE	COUNT	pH
0	5	0.0
1	5	6.0
2	3	6.2
4	2	7.5

Remap_Table		
VALUE	pH	LINK
0	0.0	10
1	6.0	50
2	6.2	100
4	7.5	75

```
slice([inraster1], table, c:\data\remap_table, value, link)
```

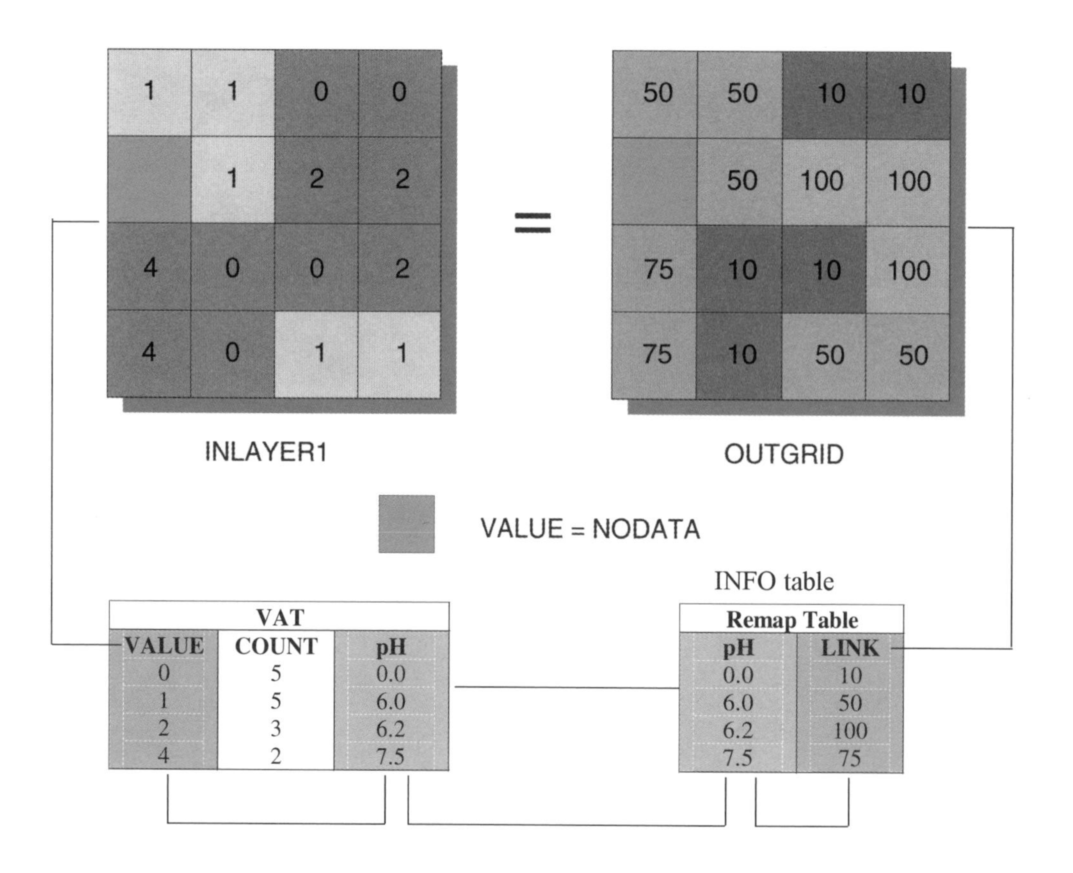


```
slice([inlayer1].ph, table, c:\data\Remap_Table, ph, link)
```

Reclass and remap tables

The Reclass function is designed for nominal data, whereas the Slice function is designed for ordinal data. The principal difference is the behavior of the Reclass function on input values that are not explicitly listed as entries in the remap table. Rather than assign such input values an output based on an inferred range, Reclass assigns them either an output of NoData or an output value identical to the input value, depending on the option selected. In the following example, a cell value is identified and then changed.

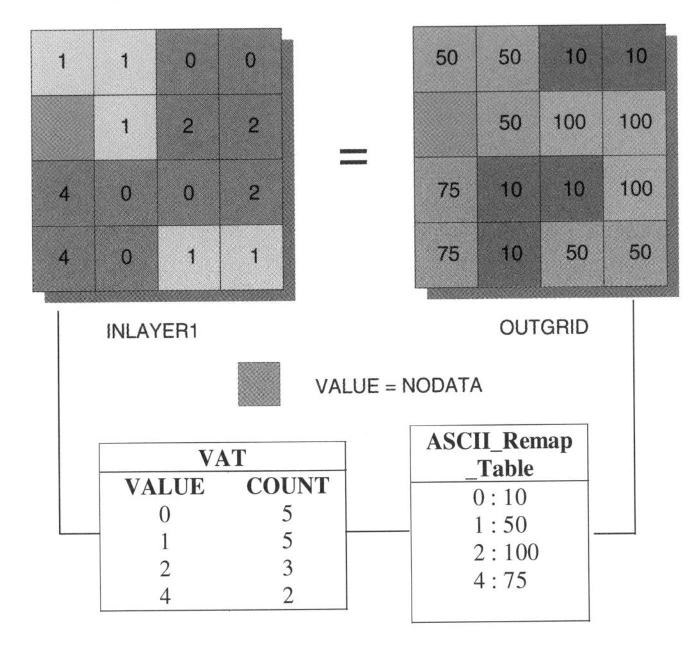


```
reclass([inlayer1], c:\data\ASCII_Remap_Table.txt)
```

Slice versus Reclass relative to remap tables

Reclass easily changes single values to alternative values, while Slice specializes in changing the ranges of values. Slice also has several special capabilities, such as dividing the cell values into groupings based on the value ranges or on the number of cells in each grouping. With certain types of reclassifications, either function can be used, and in others, only one can be used efficiently. The following list differentiates the two commands as they are used in remap tables:

- Slice always evaluates on ranges of values, while Reclass changes values singly unless the input explicitly specifies ranges.
- Reclass can copy over the original input values to the output values. In Slice, all values are affected.
- Reclass cannot specify ranges through an INFO remap table.

Glossary

altitude

1. The height, z-value, or vertical elevation of an object above a surface.
2. The angle above the horizon—measured in degrees—from which a light source illuminates a surface, used when calculating a hillshade.

analysis extent

A Spatial Analyst option to set the extent—the x,y coordinates for the bottom-left and the top-right corners—for the results from spatial analysis.

analysis mask

A Spatial Analyst option that uses a raster dataset in which all cells of interest have a value and all other cells are NoData. It enables you to perform analysis on a selected set of cells. Processing will only occur on selected cells, with other cells being assigned NoData.

arithmetic functions

Functions within the Raster Calculator of Spatial Analyst.

There are six arithmetic functions: Abs, Ceil, Floor, Int, Float, and IsNull.

The Abs function takes the absolute value of the values in an input raster.

Ceil and Floor convert floating-point values into integers by rounding up or down, respectively.

Int and Float convert values from and to integer and floating-point values.

The IsNull function returns 1 if the values on the input raster are NoData—null—and 0 if they are not.

arithmetic operators

Operators within the Raster Calculator of Spatial Analyst. They allow for the addition, subtraction, multiplication, and division of two rasters or numbers or a combination of the two.

aspect

Aspect is the direction a slope faces or the direction of steepest descent, defined by the cell and its eight surrounding neighbors.

attribute table

A table stored in rows and columns, giving information about features on a map. Each row relates to a single feature; each column contains the values for a single characteristic.

In an Integer—Categorical—raster attribute table, the first field in the table is value, which stores the value assigned to each zone of a raster. A second field, count, stores the number of cells that belong to each zone.

autocorrelation

The statistical relationship among the measured points, where the correlation depends on the distance and/or direction that separates the locations.

azimuth

The direction, measured in degrees, from which a light source illuminates a surface; used when calculating a hillshade.

barrier

A line or polygon dataset that limits the search for input sample points when performing interpolation. The line can represent a cliff, ridge, or some other interruption in the landscape. Only the sample points on the same side of the barrier as the current processing cell will be considered.

bin

A classification of lags, where all lags that have similar distance and direction are put into the same bin. Bins are commonly formed by dividing the sample area into grid cells or sectors.

Boolean operators

Operators within the Raster Calculator of Spatial Analyst. They use Boolean logic—TRUE or FALSE—on the input rasters on a cell-by-cell basis. Output values of TRUE are written as 1 and False as 0.

Boolean operators: *And*, *Or*, *Xor*, *Not*.

For example: *inraster1* or *inraster2*

Catalog tree

Contains a set of folder connections that provide access to geographic data stored in folders on local disks or shared on the network. It also includes folders that let you manage database connections and coordinate systems. The Catalog tree provides a hierarchical view of the geographical data in those folders.

categorical raster

A raster that represents the world with a set of values that have been aggregated into classes. For example, a satellite image that has been reclassified to extract a number of land cover types is a categorical raster. The cells of categorical rasters represent areas. See also discrete raster.

cell

See raster cell.

cell size

The length in map units of a side of a cell. The cell size is the same in both the x and y directions.

cell statistics

A Spatial Analyst function that calculates a statistic for each cell of an output raster that is based on the values in the same cell location of each cell of multiple input rasters.

classify

The process of sorting or arranging attribute values into groups or categories; all members of a group are represented on the map by the same symbol.

continuous raster

A raster that represents the world with a set of values that vary continuously as a surface. For example, a raster digital elevation model and an interpolated chemical concentration surface are continuous rasters. The cells of continuous rasters represent the values at the center of the cell.

contour

A contour is a line that connect points of equal value on a surface.

coordinate system

A reference system used to measure horizontal and vertical distances on a planimetric map. A coordinate system is usually defined by a map projection, a spheroid of reference, a datum, one or more standard parallels, a central meridian, and possible shifts in the x and y directions to locate x,y positions of point, line, and area features.

In ArcGIS, a system with units and characteristics defined by a map projection. A common coordinate system is used to spatially register geographic data for the same area.

cost dataset

An input dataset necessary to run the Cost Weighted Distance function using Spatial Analyst. It identifies the cost of traveling through each cell. The Cost Weighted Distance function uses this cost raster dataset to calculate the accumulative cost of traveling from every cell in the raster to a source or a set of sources.

cost weighted allocation

A Spatial Analyst function that identifies the nearest source from each cell in a cost weighted distance raster. Each cell is assigned to its nearest source cell in terms of accumulated travel cost.

cost weighted direction

A Spatial Analyst function that provides a road map from the Cost Weighted Distance raster, identifying the route to take from any cell, along the least-cost path, back to the nearest source.

cost weighted distance

A Spatial Analyst function that uses a cost raster to assign a value—the least accumulative cost of getting back to the source—to each cell of an output raster.

cut/fill

A Spatial Analyst function that summarizes areas and volumes of change between two surfaces.

data

A collection of related facts usually arranged in a particular format and gathered for a particular purpose.

data frame

A frame on the map that displays layers occupying the same geographic area. You may have one or more data frames on your map depending on how you want to organize your data. For instance, one data frame might highlight a study area, and another might provide an overview of where the study area is.

dataset

Any geographic data, such as a coverage, shapefile, raster, or geodatabase. Same as data source.

density

A Spatial Analyst function that distributes the quantity or magnitude of point or line observations over a unit of area to create a continuous raster—for example, population per square kilometer.

destination

The destination point for a path when performing the Shortest Path function.

discrete raster

A raster that typically represents phenomena that have clear boundaries and attributes that are descriptions or categories. Each cell in a discrete raster stores an integer value that represents a feature. In a raster of land cover, for example, the value 1 might represent forested land, the value 2 for urban land, and so on. See also categorical raster.

Euclidean distance

See straight line distance.

extent

The minimum rectangle bounding the area of a geographic dataset.

feature

A representation of a real-world object in a layer on a map.

feature dataset

A collection of feature classes in a geodatabase that share the same spatial reference.

field

A column in a table. Each field contains the value for a single attribute.

focal functions

This group of Spatial Analyst functions computes an output raster where the output value at each location is a function of the input cells in some specified neighborhood of the location.

format

The pattern into which data is systematically arranged for use on a computer. A file format is the specific design of how information is organized in the file. For example, raster datasets come in different formats such as ESRI grid, TIFF, and MrSID™ from LizardTech Software.

function

See spatial function or mathematical functions.

GIS

Geographic information system. An organized collection of computer hardware, software, geographic data, and personnel designed to efficiently capture, store, update, manipulate, analyze, and display all forms of geographically referenced information.

global functions

This group of Spatial Analyst functions computes an output raster where the output value at each location is potentially a function of all the cell(s) in the input raster.

grid

A geographic representation of the world as an array of equally sized square cells (or pixels) arranged in rows and columns. Each grid cell is referenced by its geographic x,y location. See raster.

hillshade

The hypothetical illumination of a surface.

histogram

A chart representing a frequency distribution. The width of each rectangle in this chart represents a class interval. The area of each rectangle is proportional to the corresponding frequency.

interpolation

A set of Spatial Analyst functions that predict values for a surface from a limited number of sample data points, creating a continuous raster.

Inverse Distance Weighted

An interpolation method where cell values are estimated by averaging the values of sample data points in the vicinity of each cell. The closer a point is to the center of the cell being estimated, the more influence, or weight, it has in the averaging process.

kriging

A surface interpolation method available in Spatial Analyst. It is a geostatistical interpolation method based on statistical models that include autocorrelation—the statistical relationship among the measured points. Kriging weights the surrounding measured values to derive a prediction for an unmeasured location. Weights are based on the distance between the measured points, the prediction location, and the overall spatial arrangement among the measured points.

lag

The line—vector—that separates any two locations. A lag has length—distance—and direction—orientation.

layer

A collection of similar geographic features—such as rivers, lakes, counties, or cities—of a particular area or place for display on a map. A layer references geographic data stored in a data source, such as a raster, coverage, or shapefile, and defines how to display it.

least-cost path

See shortest path.

local functions

This group of Spatial Analyst functions computes an output raster where the output value at each location is a function of the input value at the same location.

logarithmic functions

Perform exponential and logarithmic calculations on input rasters and numbers.

There are six logarithmic functions: base e (Exp), base 10 (Exp10), and base 2 (Exp2) exponential capabilities, and natural (Log), base 10 (Log10), and base 2 (Log2) logarithmic capabilities.

make permanent

Creates a permanent raster from a temporary analysis result.

Map Algebra

The analysis language for Spatial Analyst. It provides access to a wide range of additional functions not included in the user interface and enables you to build more complex expressions and process them as a single command.

map document

The disk-based representation of a map. Map documents can be printed or embedded into other documents. Map documents have a .mxd file extension.

map projection

See projection.

mathematical functions

Functions within the Raster Calculator that are applied to the values of a single input raster. There are four groups of mathematical functions available: Logarithmic, Arithmetic, Trigonometric, and Powers.

mathematical operators

Operators within the Raster Calculator that apply a mathematical operation to the values in two or more input rasters. There are three groups of mathematical operators available: Arithmetic, Boolean, and Relational.

Arithmetic: *, /, -, +

Boolean: And, Or, Xor, Not

Relational: ==, >, <, <>, .=, <=

model

1. An abstraction of reality.

2. A set of clearly defined analytical procedures used to derive new information.

nearest-neighbor resampling

Uses the value of the closest cell to assign a value to the output cell when resampling.

neighborhood statistics

A focal function that computes an output grid where the value at each location is a function of the input cells within a specified neighborhood of the location.

NoData

Some rasters have empty cells within the area for which data was collected. For grids, these cells are assigned NoData, while for other formats they are often assigned a special value such as -9999. Rasters with some NoData cells can also be created using the Spatial Analyst Reclassify function. You can control the display of NoData by setting the NoData color on the Symbology tab of the Layer Properties dialog box.

normalize

To make, conform to, or reduce to a standard, or norm. Data can be normalized by applying a percentage influence to each dataset, divided by 100, when performing suitability modeling. This process normalizes the values in the output suitability map to those of the input datasets.

nugget

A parameter of a covariance or semivariogram model that represents independent error, measurement error, and/or microscale variation at spatial scales that are too fine to detect. The nugget effect is seen as a discontinuity at the origin of either the covariance or semivariogram model.

operator

A mathematical symbol that performs an operation. Operators are provided in the Raster Calculator to enable analysis to be performed within and between multiple rasters. For a list of available operators, see Appendix B.

path

The location of a file or directory on a disk. A path is always specific to the computer operating system.

permanent dataset

A raster dataset permanently stored on disk. All output raster results from Spatial Analyst are temporary, unless you specify a location on disk and a filename in a function dialog box, or you make the temporary dataset permanent, or you save the map document. In these three cases, temporary results will become permanent datasets on disk. See make permanent.

pixel

See raster cell.

power functions

Apply a Power function to the values in a single input raster. Three Power functions are available: Sqrt, Sqr, or Pow.

projected coordinate system

A measurement of locations on the earth's surface expressed in a two-dimensional system that locates features based on their distance from an origin (0,0) along two axes, a horizontal x-axis representing east–west and a vertical y-axis representing north–south. A map projection transforms latitude and longitude to x,y coordinates in a projected coordinate system.

projection

A mathematical formula that transforms feature locations from the earth's curved surface to a map's flat surface. A projected coordinate system employs a projection to transform locations expressed as latitude and longitude values to x,y coordinates. Projections cause distortions in one or more of these spatial properties: distance, area, shape, and direction.

raster

Represents any data source that uses a grid structure to store geographic information. See grid.

Raster Calculator

A Spatial Analyst function that provides a powerful tool for performing multiple tasks: you can perform mathematical calculations using operators and functions, set up selection queries, or type in Map Algebra syntax.

raster cell

A discretely uniform unit, such as a square meter or a square mile, representing a portion of the earth in a raster. A cell, or pixel, has a value that corresponds to the feature or characteristics at that site, such as soil type, elevation, or landuse type.

raster dataset

Contains raster data organized into bands. Each band consists of an array of cells with optional attributes for each cell, or pixel,. Raster datasets come in many different formats. See format.

raster resolution

The size of the cells in a raster. See also cell size.

reclassify

A Spatial Analyst function that takes input cell values and replaces them with new output cell values.

region

Each group of connected cells in a zone. See zone.

relational operators

Evaluate specific relational conditions. If a condition is TRUE, the output is assigned a value of 1. If the condition is FALSE, the output is assigned a value of 0.

Relational operators: ==, <, >, <>, >=, <=.

resampling

The process of extrapolating new cell values when transforming rasters to a new coordinate space or cell size.

selected set

A subset of the features in a layer or records in a table. ArcMap provides several ways to select features and records graphically or according to their attribute values.

semivariogram

The variogram divided by two.

shapefile

A vector data storage format for storing the location, shape, and attributes of geographic features.

shortest path

A Spatial Analyst function that calculates the least-cost path from a destination point to the cheapest source using the Cost Weighted Distance and Cost Weighted Direction datasets created via the Cost Weighted Distance function.

sill

A parameter of a variogram or semivariogram model that represents a value that the variogram tends to when distances get very large. At large distances, variables become uncorrelated, so the sill of the semivariogram is equal to the variance of the random variable.

slope

Slope is the incline, or steepness, of a surface at a specific location. The slope for a cell in a raster is the steepest downhill slope of a plane defined by the cell and its eight surrounding neighbors. Slope can be measured in degrees from horizontal (0–90) or percent slope. As slope angle approaches vertical (90 degrees), the percent slope approaches infinity.

snap extent

Setting the snap extent to a specific raster will snap all layers to the cell registration of the specified raster. All layers will share the lower-left corner and cell size of the specified raster.

source dataset

A necessary input to the Cost Weighted Distance function. The source is the point or group of points that the Cost Weighted Distance function uses when calculating the accumulated cost of traveling through each cell to the nearest source.

spatial analysis

The study of the locations and shapes of geographic features and the relationships between them. Spatial analysis is useful when evaluating suitability, when making predictions, and for gaining a better understanding of how geographic features and phenomena are located and distributed.

spatial data

The locations and shapes of geometric features with descriptions of each.

spatial function

An operation that performs spatial analysis. All spatial operations on the Spatial Analyst user interface are classified as spatial functions, for example, distance, slope, or density.

spatial reference

Specifies the coordinate system of the dataset.

spline

An interpolation method where cell values are estimated using a mathematical function that minimizes overall surface curvature, resulting in a smooth surface that passes exactly through the input points.

straight line allocation

Identifies which cells belong to which source, based on closest proximity in a straight line.

straight line direction

Identifies the azimuth direction from each cell to the nearest source.

straight line distance

Calculates the distance in a straight line from every cell to the nearest source.

suitability model

A model that aids in finding optimum locations. A suitability model might identify suitable locations for a new facility or a road.

symbol

A graphic representation of an individual feature or class of features that helps identify it and distinguish it from other features.

symbology

The criteria used to determine symbols for the features in a layer. A characteristic of a feature may influence the size, color, and shape of the symbol used.

table of contents

Lists all the data frames and layers on the map and shows what the features in each layer represent.

target

The setting of the Target Layer dropdown list that determines to which layer new features will be added. The target layer is set by clicking a layer in the Target Layer dropdown list. For instance, if you set the target layer to Buildings, any features you create will be part of the Buildings layer. You must set the target layer whenever you're creating new features.

temporary dataset

A raster dataset temporarily stored on disk. All output raster results from Spatial Analyst are temporary, unless you specify a location on disk and a filename in a function dialog box, make the temporary dataset permanent, or save the map document. In these three cases, temporary results will become permanent datasets on disk. See make permanent.

trigonometric functions

Perform various trigonometric calculations on the values in an input raster.

Available trigonometric functions: Sin, Cos, Tan, Asin, Acos, Atan.

variogram

A function of the distance and direction separating two locations that is used to quantify autocorrelation. The variogram is defined as the variance of the difference between two variables at two locations. The variogram generally increases with distance and is described by nugget, sill, and range parameters.

variography

The process of estimating the theoretical variogram. It begins with exploratory data analysis, then computing the empirical variogram, binning, fitting a variogram model, and using diagnostics to assess the fitted model.

viewshed

The viewshed identifies the cells in an input surface that can be seen from one or more observation points.

working directory

A directory specified in the General tab of the Options dialog box that indicates the location on disk to place all results from analysis. All permanent and temporary results will be written here, unless otherwise specified in a function dialog box.

zonal functions

This group of functions creates an output raster in which the computation of the desired function occurs on the cell values from the input value raster that intersect or fall within each zone of a specified input zone dataset. The input zone dataset is only used to define the size, shape, and location of each zone, while the value raster is used to identify the values to be used in the evaluations within the zones.

zonal statistics

A Spatial Analyst function that calculates a statistic for each zone of a zone dataset, based on values from a raster dataset. A single output value is computed for every zone in the input zone dataset.

zone

All cells in a raster with the same value, regardless of whether they are contiguous.

Z-factor

The number of ground x,y units in one surface z unit. The input surface values are multiplied by the specified Z-factor to adjust the input surface z units to another unit of measure.

Index

A

B

C

D

E

F

G

H

I

K

L

S

T

V

W

Z